Synthesis Lectures on Digital Circuits & Systems

Series Editor

Mitchell A. Thornton, Southern Methodist University, Dallas, USA

This series includes titles of interest to students, professionals, and researchers in the area of design and analysis of digital circuits and systems. Each Lecture is self-contained and focuses on the background information required to understand the subject matter and practical case studies that illustrate applications. The format of a Lecture is structured such that each will be devoted to a specific topic in digital circuits and systems rather than a larger overview of several topics such as that found in a comprehensive handbook. The Lectures cover both well-established areas as well as newly developed or emerging material in digital circuits and systems design and analysis.

Zain Ul Abideen · Samuel Pagliarini

Reconfigurable Obfuscation Techniques for the IC Supply Chain

Using FPGA-Like Schemes for Protection of Intellectual Property

Springer

Zain Ul Abideen
Carnegie Mellon University
Pittsburgh, PA, USA

Samuel Pagliarini
Carnegie Mellon University
Pittsburgh, PA, USA

ISSN 1932-3166 ISSN 1932-3174 (electronic)
Synthesis Lectures on Digital Circuits & Systems
ISBN 978-3-031-77511-6 ISBN 978-3-031-77509-3 (eBook)
https://doi.org/10.1007/978-3-031-77509-3

This Springer imprint is published by the registered company Springer Nature Switzerland AG
The registered company address is: Gewerbestrasse 11, 6330 Cham, Switzerland

With the name of Almighty Allah, Zain Ul Abideen would like to dedicate this book to his family, friends and teachers.

Samuel Pagliarini would like to dedicate this book to all his past and current students and the incredible contributions they have made to the field of Hardware Security.

Foreword

Innovations in the design and fabrication of Integrated Circuits (ICs) have profoundly changed our lives and the way we interact, and rely on, computing devices. Applications and functions that are only possible via IC-based systems are ubiquitous. We now have cellphones with more computing power than that of spacecraft that went to the moon and back. We have cars that are operated using the hundreds of chips within them, automating all aspects of the vehicle, from the driving itself to the passenger entertainment.

One side effect of this chip industry evolution over the past few decades is that the supply chain has become incredibly complex and distributed. For a chip to be conceived, the typical case is that multiple entities are involved. Many leading electronics companies now design their products in-house but outsource the manufacturing, testing, and packaging to specialized vendors. While this is a cost-effective measure, it creates the potential for malicious behavior in the form of reverse engineering. In principle, determined adversaries can analyze and replicate electronic components that they get their hands on. To address this, protecting the Intellectual Property (IP) that resides within a chip is crucial.

Numerous researchers, both in academia and government, have explored approaches for protecting chip IP. Logic locking approaches were proposed whereby a design only operates correctly when provided with an authorized key value. Others have explored hybrid manufacturing approaches where part of the design is fabricated in a trusted way. Recently, the emergence of reconfigurable-based obfuscation (ReBO) has changed the landscape in IP protection, providing a solution that is FPGA-inspired and likely more apt to combat threats to IP integrity. eFPGA redaction, for example, is a promising technique in the ReBO space.

The authors of this book have laid the foundation for exploring design versus security trade-offs for ReBO. The authors provide a chronological overview of the evolution of FPGA-like IP obfuscation techniques and describe a comprehensive threat model, security metrics, and potential attacks, both existing and future. At its core, this book provides a footprint for a ReBO design methodology to protect chips of all scale and complexity against current and predicted threats to their IP integrity.

Given the authors' experience in developing ReBO defenses and understanding offensive strategies, I believe this book will be a valuable resource for both academic researchers and industry professionals.

September 2024

Larry Pileggi
Coraluppi Head and Tanoto Professor
Department of Electrical and Computer Engineering
Carnegie Mellon University
Pittsburgh, PA, USA

Preface

The increasing complexity of integrated circuits and the rising expenses of owning and operating foundries have given rise to the fabless business model. Fabless companies, i.e., design houses, focus on product design and marketing while outsourcing the fabrication, test, and assembly processes to specialized offshore foundries and test/assembly companies. However, this outsourcing to globalized entities has increased the risk of security threats such as IP piracy, overbuilding, counterfeiting, and reverse engineering, which can pose significant economic and safety risks.

To address these challenges, various countermeasures have been developed, including watermarking, split manufacturing, camouflaging, hardware metering, and Logic Locking (LL). LL has been in existence for over a decade, but the current status of this technique is a never-ending "cat and mouse" game between defenses and attacks, which is still ongoing. LL involves integrating additional logic into a circuit during the design phase and giving it a key-dependent operation. In the last few years, reconfigurable-based obfuscation (ReBO) emerged as a promising and versatile alternative to LL and the other aforementioned countermeasures.

The concept of ReBO has gained significant attention, with ongoing advancements and a growing body of research. Researchers and industry professionals are increasingly focused on developing stronger defenses and understanding various potential attack methods. The authors of this book have been actively involved in ReBO research, contributing papers, and participating in the development of computer-aided design (CAD) tools. This expanding interest has led to this book, an effort to consolidate the knowledge behind ReBO. This book aims to describe the intricacies of obfuscation methods that leverage the flexibility and adaptability of reconfigurable logic. Scope-wise, this book provides a comprehensive overview of reconfigurable-based obfuscation, outlining core principles and emphasizing key research findings. Consistent with research developments, the majority of the techniques require thorough security analysis. The material organizes knowledge pertaining to defenses and classification of obfuscation techniques, providing detailed explanations of attacks and highlighting essential distinctions between them.

Overall, the book is divided into two parts. Part I describes various characteristics of the ReBO concept, tracing the evolution of ReBO techniques over time. It provides a detailed classification of these techniques and offers an in-depth description of the three major classes. This part also describes the novel attacks targeted at the ReBO techniques. Concerning the three major classes, this part describes the practical demonstration of each class, followed by a security analysis.

In Part II of the book, we focus on the hybrid ASIC (hASIC) solution that we have developed over the years. This solution emphasizes the concept of design versus security trade-offs. We have developed a custom tool that facilitates security-aware obfuscation compatible with standard cell-based physical synthesis. The tool explores the FPGA-ASIC design space to generate hybrid ASICs, which maintain high security while delivering ASIC-like performance. We present various experiments to demonstrate how a designer can leverage the features to make informed decisions on an intricate trade-off space. Additionally, this part presents the results from physical synthesis, layouts, novel security analysis, and discussions.

The book is organized in a way that provides clarity and context, offering chapter summaries at the beginning of each part to help readers navigate through the content. The intended audience for this book includes senior and graduate students in electrical and computer engineering, as well as professionals in IC design and CAD software development who have a basic understanding of the IC design flow. It serves various purposes such as being a textbook for the intersection of hardware security with VLSI CAD and IC design. It also functions as a comprehensive "designer's guide" for implementing different ReBO techniques in ICs. The book introduces the foundational concepts of ReBO in a systematic manner, making it accessible even to those who are new to the field.

Pittsburgh, PA, USA
August 2024

Zain Ul Abideen
Samuel Pagliarini

Acknowledgements

The authors express their gratitude to the Department of Electrical and Computer Engineering at Carnegie Mellon University in the USA, as well as to the Department of Computer Systems and the School of IT staff at Tallinn University of Technology in Estonia. They also extend their thanks to their colleagues Dr. Levent Aksoy, Dr. Muayad Baqer Al-Jafar, Dr. Malik Imran, Dr. Tiago Perez, and Felipe Almeida. Special appreciation is also extended to Dr. Mohammad Eslami and Carlos Gabriel de Araujo Gewehr for their valuable proofreading contributions. This book's research was made possible in part by the support of the Estonian Research Council through the MOBERC35 project, the European Commission through the SAFEST project, the IT Academy, the European Social Fund, and the Estonian Education and Youth Board through the EITSA18019 project, as well as through the SMART4ALL program under the EU's Horizon 2020 R&D initiative.

Contents

Acronyms

3PIP	Third-Party Intellectual Property
ABEL	Advanced Boolean Expression Language
AES	Advanced Encryption Standard
AI	Artificial Intelligence
ASIC	Application-Specific Integrated Circuit
AT	Arrival Time
ATPG	Automatic Test Pattern Generation
BCHD	Between-Class Hamming Distance
BEOL	Back-End-Of-the-Line
CAD	Computer-Aided Design
CGRRA	Coarse-Grained Runtime Reconfigurable Array
CLB	Configurable Logic Block
CMOS	Complementary Metal-Oxide Semiconductor
CPU	Central Processing Unit
CTS	Clock Tree Synthesis
DIP	Dual-In-Line
DP	Double Pattern
DPA	Differential Power Analysis
DRAM	Dynamic Random Access Memory
DRC	Design Rule Check
DSP	Digital Signal Processing
DUO	Design Under Obfuscation
EDA	Electronic Design Automation
eFPGA	Embedded-Field Programmable Gate Array
EUIPO	European Union Intellectual Property Office
FC	Functional Composition
FEOL	Front-End-Of-the-Line
FFT	Fast Fourier Transform

FPGA	Field-Programmable Gate Array
FPU	Floating Point Unit
GDS	Graphic Data System
GE	Gate Equivalent
GPU	Graphics Processing Unit
GSHE	Giant Spin Hall Effect
hASIC	Hybrid ASIC
HDL	Hardware Description Language
HPC	High Performance Computing
HVT	High Voltage Threshold
HW	Hamming Weight
IC	Integrated Circuits
IIR	Infinite Impulse Response
IoT	Internet of Things
IP	Intellectual Property
ITU	International Telecommunication Union
LEF	Liberty Exchange Format
LL	Logic Locking
LSI	Large-Scale Integration
LUT	Look-Up Table
LVT	Low Voltage Threshold
MESO	Magneto-Electric Spin-Orbit
MHW	Masked Hamming Weight
MIPS	Microprocessor without Interlocked Pipelined Stages
MLP	Machine Learning Processing
MRAM	Magnetic-Random Access Memory
MTJ	Magnetic Tunnel Junction
NIST	National Institute of Standards and Technology
NoC	Network on Chip
NVM	Non-Volatile Memory
P&R	Place & Route
PCB	Printed Circuit Board
PDK	Process Design Kit
PID	Proportional Integral Derivative
PLL	Phase-Locked Loop
PPA	Power-Performance-Area
PSCA	Power Side-Channel Attacks
PUF	Physical Unclonable Function
QoR	Quality of Results
QP	Quad Patterning
RAM	Random Access Memory

RE	Reverse Engineering
RISC	Reduced Instruction Set Computer
RSA	Rivest–Shamir–Adleman
RT	Required Time
RTL	Register-Transfer Level
SAT	Satisfiability
SBM	Schoolbook Multiplier
SDC	Synopsys Design Constraints
SHA	Secure Hash Algorithm
SMIC	Semiconductor Manufacturing International Corporation
SoC	System on Chip
SOT	Spin-Orbit Torque
SP	Single Pattern
SRAM	Static Random Access Memory
STA	Static Timing Analysis
STT	Spin-Transfer Torque
SVT	Standard Voltage Threshold
TNS	Total Negative Slack
TOTe	Tunable design Obfuscation Technique
TP	Tripple Pattern
TPU	Tensor Processing Unit
TRAP	TRAnsistor-level Programming
TSMC	Taiwan Semiconductor Manufacturing Company
ULSI	Ultra Large-Scale Integration
VLSI	Very Large-Scale Integration
WCHD	With-in Class Hamming Distance
WCSHD	With-in Class Sequential Hamming Distance
WNS	Worst Negative Slack

Part I
Overview and Security in Reconfigurable-Based IC Design

Part I of this book introduces ReBO as a cutting-edge solution to enhance security in the globalized integrated circuit (IC) supply chain. ReBO leverages reconfigurable hardware elements to secure the functionality of a design until it is programmed post-fabrication, providing robust protection against various hardware-based attacks. We describe the critical role of ReBO and its integration into the IC design flow, covering technological advancements, security threats, countermeasures, classifications, and much more.

Chapter 1 discusses the globalized IC supply chain's significance and the increasing importance of hardware security. It outlines the threats posed by globalization, such as reverse engineering and intellectual property (IP) piracy, and reviews existing security measures. The limitations of these measures are highlighted, introducing ReBO as a novel approach to counter hardware-based attacks by concealing design functionality behind a programmed bitstream.

Chapter 2 focuses on field-programmable gate arrays (FPGAs) and embedded FPGAs (eFPGAs); exploring their architecture, programming methodologies, and applications. It emphasizes the flexibility and performance optimization offered by ReBO, setting the foundation for understanding ReBO's role in enhancing security and adaptability in IC design.

Chapter 3 provides a comprehensive overview of IC design, particularly when incorporating ReBO. It explains how ReBO offers obfuscation to secure chips conceived within the globalized IC supply chain. It also compares it with traditional countermeasures and LL techniques. The chapter discusses the vulnerabilities in existing methods and how ReBO strengthens security against supply chain threats, highlighting the need for custom CAD tools and addressing end-user stage security concerns.

ReBO's techniques and implementations are explored in depth, focusing on the use of reconfigurable elements like SRAM-based LUTs and non-volatile memory (NVM)-based LUTs. Chapter 4 discusses advanced technologies such as spin-transfer torque (STT), magnetic tunnel junction (MTJ), and spin-orbit torque (SOT), providing a comprehensive classification of ReBO approaches based on technology, element type, and IP type. It evaluates the trade-offs between security and power, performance, and area (PPA), offering detailed comparisons of different ReBO methods.

Chapter 5 underscores the importance of security analysis for ReBO, discussing various attack types and security metrics, such as output corruptibility and execution time.

It covers oracle-guided and oracle-less adversarial settings, including the SAT attack and its variants that specifically target ReBO. The chapter also highlights the need for extensive security analysis of existing technologies and evaluates different ReBO techniques, emphasizing the necessity for improved security measures.

LUT-based obfuscation, a central aspect of ReBO, is examined in detail. Chapter 6 describes various methods proposed by researchers, discussing the selection of design parts for reconfigurable and non-reconfigurable (static) segments and the associated security versus PPA trade-offs. It reviews approaches, their security against the SAT attack, and common pitfalls due to inadequate security analysis. Chapter 6 also covers heuristic selection methods, logic synthesis-inspired techniques, logarithmic-based network obfuscation, and the first silicon demonstration, concluding with a comparative analysis.

Chapter 7 focuses on eFPGA redaction, detailing methods and techniques used to enhance security through selective obfuscation. It highlights the practical implementations and benefits of integrating eFPGA technology within an IC and provides insights into its role in ReBO.

Chapter 8 explores hybrid approaches that combine various CMOS and emerging technologies. It discusses the advantages of emerging technologies in achieving robust security and performance optimization, offering a comprehensive understanding of their role in modern IC design and ReBO's contribution to these advancements.

By introducing ReBO and examining its application across different chapters, Part I provides a thorough understanding of how ReBO enhances security in IC design, addressing the challenges posed by the globalized IC supply chain.

Introduction

1

1.1 Demand and History of High-Performance ICs

The digitalization of critical infrastructure has become increasingly important in modern society [1]. Critical infrastructure refers to the systems and assets essential for a country's economy, security, and public health, such as transportation networks, energy grids, water supply systems, and healthcare facilities [2]. The goal of digitalizing critical infrastructure is to make human tasks more efficient, convenient, and faster, which can significantly impact various aspects of our daily lives [3, 4]. The process of digitalization critical infrastructure is complex and multifaceted. Nowadays, technological advances such as artificial intelligence (AI), the internet of things (IoT) paradigm, as well as edge computing [5, 6] are practical and being readily utilized.

IC-based systems are pivotal components in technological advances, playing a crucial role in the digitalization of critical infrastructure. In many ways, the performance of a critical system, as perceived by its users and stakeholders, is a function of the performance of the ICs that comprise it. Achieving high performance in ICs requires fabrication on advanced technology nodes, enabling rapid information processing, lower power consumption, and higher transistor densities on a chip. IC foundries continuously refine their fabrication processes (lithography, deposition, patterning, etc.) to meet these evolving demands.

1.1.1 History of the IC

The introduction of the transistor by Bell Labs scientists John Bardeen, Walter Brattain, and William Shockley in 1947 was a significant milestone in electronics [7]. It replaced the bulky and unreliable vacuum tube commonly employed in electronic devices at the time with a smaller, more reliable, and power-efficient alternative. This breakthrough soon led to

Z. U. Abideen and S. Pagliarini, *Reconfigurable Obfuscation Techniques for the IC Supply Chain*, Synthesis Lectures on Digital Circuits & Systems,
https://doi.org/10.1007/978-3-031-77509-3_1

the development of ICs, which were created by integrating multiple transistors on a single chip [8], as the name implies. In addition to transistors that are active devices, an IC also includes passive devices such as capacitors and resistors, all integrated into a single, compact package. The first IC developed by Jack Kilby at Texas Instruments in 1958 only contained a few transistors. In 1961, the world's first commercial IC [9], the N51x series, was released, demonstrating great potential to revolutionize electronics, as shown in Fig. 1.1. Figure 1.2 illustrates the evolution of computer chips and their transistor counts. In the early 1950s, it was challenging to integrate many transistors on a single chip. However, by the early 1960s, dozens of transistors could be integrated into a single chip, leading to medium-scale integration (MSI) of circuits. By the end of the 1960s, designers could integrate hundreds of

Fig. 1.1 The SN514 IC released by Texas instruments [10]

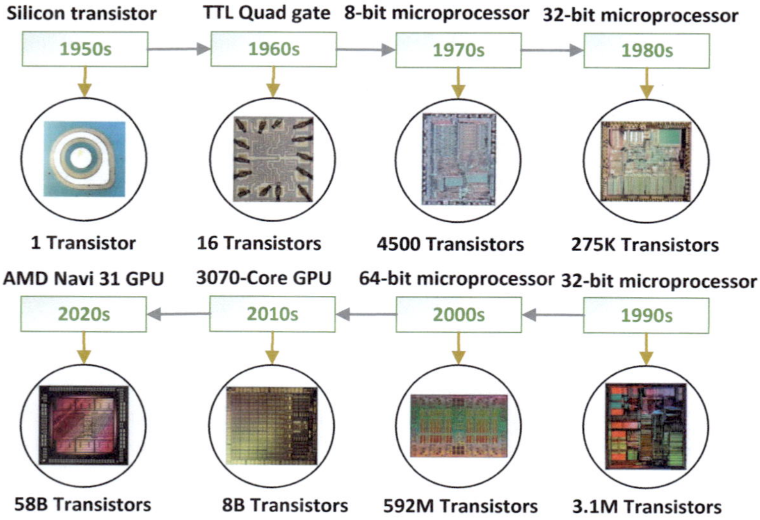

Fig. 1.2 Evolution of the number of transistors in a single chip over time [12]

transistors, leading to large-scale integration (LSI). By integrating more and more electronic components onto a single chip, designers can reduce the size of electronic devices while improving their performance and reducing power consumption. By the 1970s, thousands of transistors were being integrated onto a single chip, resulting in the very large-scale integration (VLSI) of circuits.

The advances in IC fabrication and miniaturization have led to the development of highly sophisticated and portable personal computers. Towards the end of the 1980s, the number of transistors on a single chip had increased from thousands to millions, and the technology was then dubbed ultra-large-scale integration (ULSI). The era of ULSI began with Intel's i860 chip, a RISC processor [11]. In the 2000s, this number increased to hundreds of millions of transistors on a chip, resulting in the development of 64-bit microprocessors for personal computers. Advancements in transistor technology have enabled an impressive increase in the number of transistors that can be integrated into a single chip [12]. The need for more compact, powerful, and efficient electronic devices drove the progression towards LSI, VLSI, and USLI [13]. The evolution of ICs has been driven by advances in semiconductor technology, which have enabled the creation of increasingly complex circuits. In the 2010s, remarkable technological advances led to the integration of billions of transistors onto a single chip. This trend is expected to continue with the development of new materials, fabrication techniques, and design methodologies, leading to even more advanced ICs and systems composed of ICs. The principle of Moore's law states that the number of transistors in an IC doubles approximately every two years. Surprisingly, this forecast remains relevant to this day.

1.1.2 Technology Nodes and Miniaturization

The rapid miniaturization of ICs has enabled the development of more powerful and efficient electronic devices. The semiconductor industry has undergone significant changes over the years until it assumed the globalized structure it has today. In the 1980s, Japan dominated the semiconductor market due to its superior fabrication processes, providing better yield [14]. In the 1990s, the semiconductor market experienced a significant shift as emerging economies like Korea and Taiwan began to dominate the industry. The IC foundries in these countries made substantial investments, primarily focusing on their fabrication processes. Notably, they were known for their exceptional commitment to capital expenditure, often reinvesting 100% of their revenue back into researching better materials and techniques for chip fabrication. This strategic approach allowed Korea and Taiwan to rapidly expand their semiconductor fabrication capabilities and gain a competitive edge in the global market [15]. By consistently allocating a significant portion of their resources towards capital expenditure, they enhanced their production capacity, upgraded equipment and technologies, and improved overall efficiency.

The most significant shift occurred as companies with advanced fabrication processes adopted a specialized approach. This approach involved offering the service of pure-play foundries, i.e., foundries that would focus solely on the fabrication of semiconductors [16]. This marked a departure from the vertical business model that prevailed in the early years of the semiconductor business.

Pure-play foundries offered various benefits with respect to other semiconductor companies by specializing in fabrication and in fabrication only. They provided access to state-of-the-art transistors and extensive production capacity/volume. Design companies could outsource their fabrication needs to these foundries, allowing them to focus on research, design, and marketing aspects. This trend enabled semiconductor companies to optimize resources, reduce capital expenditure, and enhance flexibility in meeting market demands. It also allowed smaller semiconductor companies to access cutting-edge fabrication technologies without significant upfront investments in fabrication facilities. With the advent of globalization and the rise of new players in the market, the semiconductor supply chain has truly become globalized, complex, and diversified.

The technology landscape of the semiconductor industry has undergone significant changes over the years [17], as depicted in Fig. 1.3. The number of cutting-edge fabrication facilities available globally is decreasing, while the older technologies such as 130nm and 90nm are still operational and remain relatively available. Building and maintaining semiconductor fabrication facilities have become prohibitively expensive for many companies [18]. This has led to consolidation in the industry, with fewer players capable of affording the significant capital expenditures needed to stay at the cutting edge of semicon-

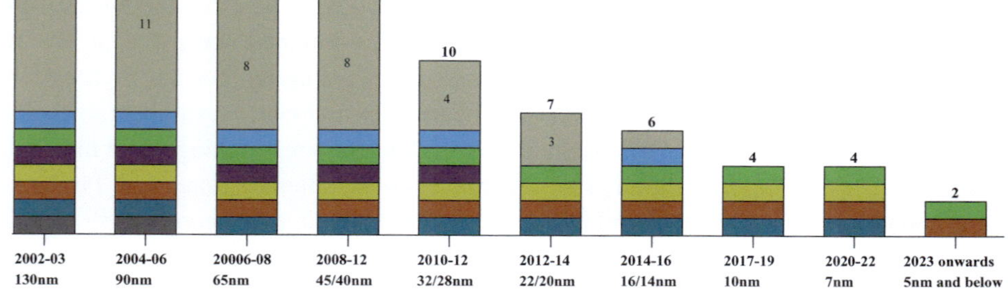

Fig. 1.3 The semiconductor industry evolution up to 10 nm [19]

ductor technology. The semiconductor industry is predominantly led by three major players: Intel, Samsung, and TSMC, as shown in Fig. 1.3.

From 2017 to 2019, the fabrication industry witnessed the emergence of the 10nm technology node. In early 2020, an additional player emerged as semiconductor manufacturing international corporation (SMIC), a pure-play foundry [20] based in China. SMIC announced its fabrication on the 7nm technology node, *allegedly* joining the ranks of Intel, Samsung, and TSMC as critical players in this advanced node [21]. However, investigations are ongoing regarding the illegal use of US technology in SMIC's 7nm technology node [22]. This development showcases the high competition levels in the semiconductor industry and the growing interest in pushing the boundaries of process technology. These companies have the financial resources and technical expertise to invest in advanced fabrication processes and push the boundaries of chip fabrication. In addition to the high fabrication costs, the semiconductor market also witnesses significant R&D expenditures. Companies allocate substantial resources to research and development activities to drive innovation, improve fabrication processes, and meet the demands of emerging technologies and applications [23]. These R&D investments are essential for maintaining competitiveness and staying ahead in the highly dynamic semiconductor industry.

Pursuing denser and faster ICs has led to a significant increase in the complexity of the fabrication process itself [24]. On the design side, the increasing complexity has led to the adoption of advanced EDA tools, customized and varied IP libraries, and innovative implementation techniques. Advanced packaging techniques, stacked die technologies, and other assembly methods have become essential to meet the demands of advanced systems [25]. In order for a design house to produce a marketable modern IC, it must secure access to process design kits (PDKs) from a foundry, license agreements with capable EDA tool vendors, and partnerships or royalty agreements with specialized IP providers. Even industry giants like Intel, who control their fabrication processes, eventually seek assistance from external entities to develop their products. The complexity of device fabrication and testing has experienced significant growth, particularly as the industry transitioned to advanced technology nodes, such as 10nm and 7nm and beyond. The challenges in printing intricate design patterns and conducting thorough device testing have increased exponentially with each new node, as illustrated in Fig. 1.4.

The complexity of conceiving an IC is also reflected in the number of design rules and fabrication steps involved in the fabrication process. Advanced technology nodes have witnessed an exponential growth in design rules, indicating the increasing intricacy of IC design [28]. The complexity of masks or patterns differs from one technology to another, as shown in Fig. 1.4. Concerning advanced technology nodes, extreme ultraviolet (EUV) lithography technology is showing potential in reducing design and manufacturing complexity. However, the deployment has been limited due to the complexity, technological immaturity, and cost of EUV equipment for 22 nm and below. Many foundries use multi-patterning to accurately print features in higher density on the growing number of silicon layers, even while utilizing EUV on certain layers.

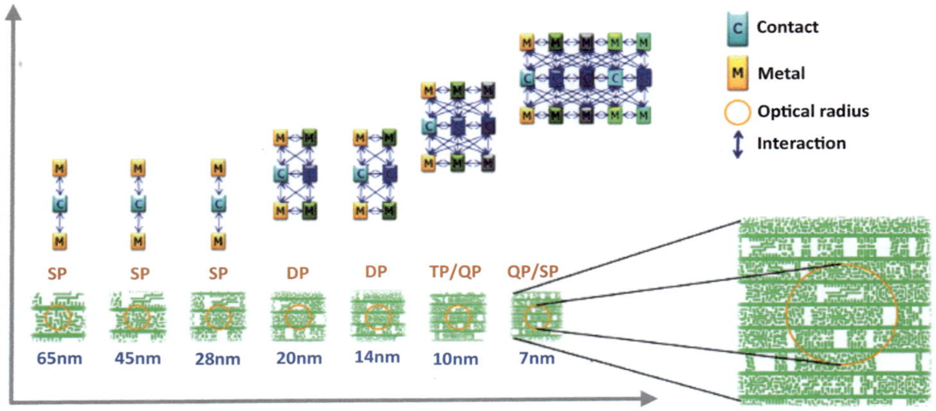

Fig. 1.4 The complexity of device fabrication on the advanced technology nodes [26, 27]

The analysis highlights two notable trends in patterning: the interplay of layers and the density of features within the optical radius, which is the influential zone typically taken into account during optical proximity correction (OPC), and the use of resolution enhancement techniques (RET). The optical radius, around 0.6um, is approximately three times larger than the wavelength of the scanner (193nm), a constant over multiple process generations. Fabrication facilities typically use Single Pattern (SP) for the 65–28 nm technologies, double pattern (DP) for 20–14 nm, and tripple pattern (TP) and quad pattern (QP) are utilized (or considered) for more advanced technologies [26]. The complexity of fabricating advanced ICs underscores the need for a collaborative ecosystem. These collaborations enable companies to leverage specialized expertise, access state-of-the-art fabrication processes, and effectively tackle the challenges posed by the growing complexity of IC design and fabrication.

In a nutshell, accessing specialized equipment and advanced fabrication processes available only in a few foundries presents particular challenges. Therefore, building and maintaining a foundry becomes more complex as the industry evolves, resulting in rising costs [29]. For instance, an estimated USD 33-34B would be required to build a foundry with 2 nm capabilities [30]. Thus, virtually every semiconductor company is following the globalized IC supply chain.

1.2 Globalization of the IC Supply Chain

This section contrasts the traditional IC design flow and fabrication process with its counterpart in a globalized supply chain. It is important to note that only a basic understanding of the IC design flow is necessary to grasp the ideas discussed in this chapter.

1.2.1 Conventional IC Design and Fabrication

The conventional IC design is straightforward and, in principle, has no security concerns. Figure 1.5 illustrates the conventional IC design flow and the subsequent fabrication where every individual stage is executed in a trusted environment. In other words, all steps are considered to be executed in-house or in a 'trusted' setting.

The conventional IC design begins with the creation of system specifications. These specifications are then translated through a series of logical design steps to generate a netlist (schematic) comprising interconnected logic gates, each representing a Boolean function. Subsequently, the netlist undergoes a series of physical design steps to map out all the gates and interconnects into a geometric layout. This involves block-level implementations and then organizing these blocks and interconnects with the help of EDA tools. Often, design houses develop their own blocks to target the specific design [31]. After developing certain blocks internally, they are combined with 3PIPs to form the overall system. EDA tools utilize the gate-level netlists to generate a layout. The layout includes instances of metal shapes, contacts, vias, and transistors that together determine the function of the chip. The layout is then sent to the foundry utilizing a file format known as graphic data stream (GDS).

A foundry transforms the given layout into masks created for the photolithography of silicon wafers, leading to the production of bare dies. Once the dies have been fabricated, they are packaged. This involves enclosing the IC in a protective casing that shields it from external elements and provides electrical connections to the device. Following fabrication, the ICs are extensively tested for manufacturing defects, and ICs deemed functional are subsequently distributed to end-users for integration into various systems.

In the past, major semiconductor companies like Intel, Texas Instruments, IBM, and many others typically utilized a vertically integrated business model, wherein they controlled the entire IC design and manufacturing process in-house, including ownership of foundries and trusted environment for design, fabrication, and testing [19]. The scenario has evolved over time with the rise of the globalized IC supply chain. Semiconductor companies now have limited control over the entire process, from design to fabrication and other stages. A few

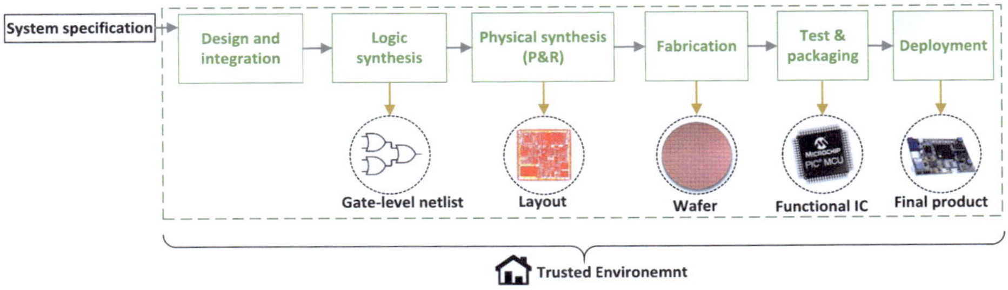

Fig. 1.5 Traditional IC design and fabrication: all stages are happening in a trusted setting

foundries with advanced technology, such as Intel and Samsung, serve as recent examples with their own fabrication facilities and resources extending to deployment stages [32].

1.2.2 Globalized IC Design and Fabrication

To be competitive in the high-performance IC business, design houses increasingly rely on globalized IC supply chains. For example, Apple will outsource the fabrication of their processor chips on 3nm from TSMC [33]. Adopting a globalized IC supply chain offers design houses the advantage of accessing high-end semiconductor facilities [34] without upfront costs. It has become common practice for various entities, including corporations and governments, to contract IC fabrication to third-party foundries. The globalized IC supply chain involves numerous dependent and interdependent tasks that can be nearly hectic to manage. Figure 1.6 illustrates the primary stages of the globalized IC supply chain. Considering the globalized IC design methodology, many stages illustrated in Fig. 1.6 differs from the traditional design methodology. The significant stages are highlighted in red.

The production of ICs involves, at a minimum, collaboration with a third-party foundry responsible for fabrication. Additionally, testing and packaging of ICs are typically outsourced to specialized third-party companies. Stringent testing ensures that the ICs meet required performance and quality standards. The deployment of ICs in products also occurs within the untrusted facility, forming a pivotal component of the global IC supply chain. In short, designing an IC and all other post-design steps that follow it involve multiple entities in the globalized IC supply chain. The green color in Fig. 1.6 visually indicates the typically trustworthy steps in the design flow. The color red indicates the untrusted stages in the globalized IC supply chain. The design layout is inevitably exposed to at least one untrusted entity. This does pose varied security risks, as illustrated in Fig. 1.6. While all entities involved provide reasonable assurances, it is essential to acknowledge the poten-

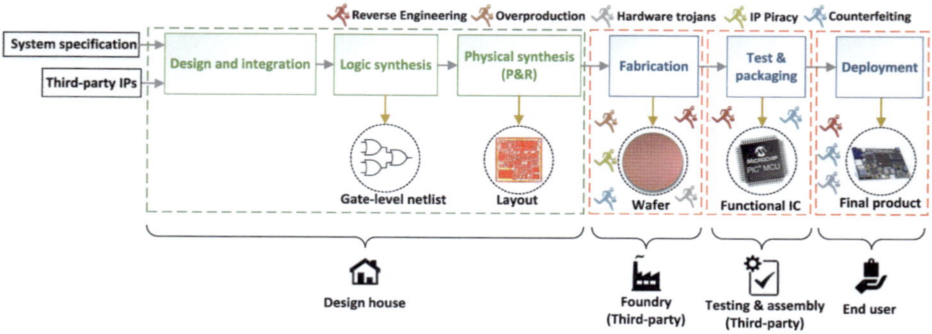

Fig. 1.6 Typical stages involved in the globalized IC supply chain: untrusted stages are highlighted in red

tial lack of a 100% guarantee concerning their trustworthiness and integrity. This lack of a guarantee, primarily stemming from the zero-trust concept, means that the foundry and its employees may pose adversarial threats. Fabrication holds the most significant importance among the various stages because the foundry can access detailed design information at a very low level.

The potential consequences of these security threats can be severe, including service interruptions, compromised public data integrity, and financial losses. Notably, both the European Union (EU) and the United States have issued warnings about the national security risks associated with scammers taking advantage of the IC supply chain crunch [35]. The many *security threats* in the globalized IC supply chain are discussed in the following section.

1.3 Hardware Security Threats

In a globalized and distributed IC supply chain, numerous entities with varying levels of trust may have access to valuable IP or the physical IC. This increased accessibility to critical assets creates opportunities for untrusted entities to exploit valuable information and potentially undermine trust in the IC design process [34, 36]. As depicted in Fig. 1.6, various security threats are recognized, such as reverse engineering (RE), overproduction, hardware trojans, IP piracy, counterfeit ICs, fault-injection attacks, and side-channel attacks [36].

1.3.1 Reverse Engineering

Two types of RE apply to ICs: physical RE and logical RE. Physical RE is accomplished through various imaging tools and methods. This process involves intricate steps, including removing the package of an IC, delayering, performing image analysis, alignment, and stitching to reconstruct the design's netlist [37], as illustrated in Fig. 1.7.

RE is often associated with malicious goals such as IP piracy, malicious logic insertion and extracting sensitive information, including cryptographic keys. Yet, RE can also be a tool for identifying vulnerabilities. Despite the difficulties involved in RE, determined adversaries can still perform it given a reasonable level of resources is allocated to the task.

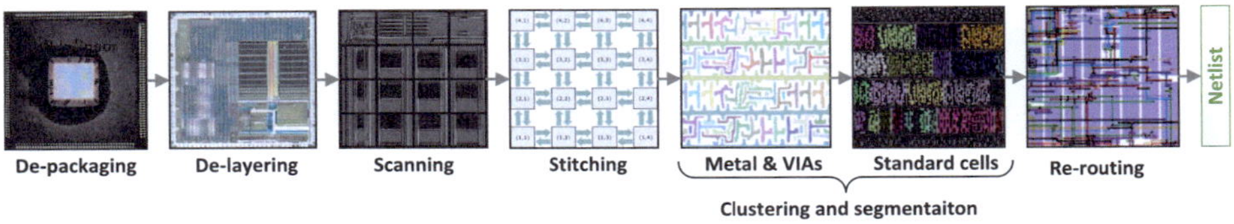

Fig. 1.7 Different steps involved in the conventional RE

Logic RE, on the other hand, is the process of making sense of the design, assuming the layout has already been converted to a gate-level netlist [38]. Logical RE involves employing techniques like structural and statistical analysis to comprehend the operation of an IC and retrieve its modules, such as finite state machines (FSMs) and register grouping [39]. Logical reverse engineering can also be utilized to recover the complete functionality of the circuit, including methods to analyze the data flow of flip-flops (FFs) [40]. This also applies to designs that contain FSMs, where the adversary aims to track the states and obtain the full functionality of the FSM [41].

1.3.2 Overproduction

Overproduction occurs when the foundry fabricates more ICs than contracted. Untrusted foundries may be motivated to overproduce ICs and distribute them at lower prices in the grey or black market [42]. Overproduction is cost-effective as foundries can use the same set of masks to produce both original and counterfeit ICs.

1.3.3 Hardware Trojans

Hardware trojans describe malicious modifications to an IC. Such modifications may be the addition, removal, or substitution of parts of the original logic of the IC. Trojans can also be parametric, such that they affect the reliable operation of the IC over its lifetime [43, 44].

Additive hardware trojans are characterized by intricate malicious logic hidden behind a stealthy trigger. Such trojans are designed to disrupt the normal operations of the IC or extract sensitive data [45, 46]. It should be noted that trojans (or backdoors) embedded in 3PIPs are also of concern since these may include hidden functionalities that reveal restricted design aspects or extract confidential information. Detecting and identifying trojans can be challenging due to their small size with respect to the entire IC, as well as the lack of a reference or “golden” design for cross-validation. With extensive access to the layout, the foundry can determine potential locations for trojan insertion [47] that minimize the detection probability.

1.3.4 IP Piracy

IP piracy occurs when the IPs used in a design are unlawfully obtained. Untrusted foundries may be interested in the unauthorized use, reproduction, and distribution of the IP of others. In a foundry environment, unauthorized individuals may engage in illicit activities, such as stealing valuable information through RE or selling IPs without proper authorization from the owner. The same concerns exist post-fabrication when an end-user can also engage in

RE for IP theft. Concerning losses associated with IP piracy, it is estimated that the US experiences an annual loss of up to USD 600B due to IP piracy [48].

1.3.5 Counterfeiting

Counterfeit ICs are unauthorized replicas that intentionally resemble genuine ICs, exhibiting similar or identical functionality. These ICs are less reliable and/or have degraded performance. Unauthorized companies or unethical sellers often supply these ICs for use in electronic products in an almost oblivious way. Counterfeit ICs can be classified into seven types such as recycled, remarked, out-of-spec/defective, cloned, forged documentation, and tampering, as shown in Fig. 3 of [49]. High-value ICs, such as FPGA boards and GPU cards, are typical targets of recycling and remarking. Small pervasive analog ICs, such as converters, are the typical cloned ICs.

In 2015, the International Telecommunication Union (ITU) and the European Union Intellectual Property Office (EUIPO) have jointly disclosed that counterfeit electronics were responsible for a 12.9% reduction in legit smartphone sales. This resulted in a substantial monetary loss of EUR 45.3B for legitimate industries [50].

1.3.6 Fault-Injection Attacks

Fault injection attacks encompass a wide range of techniques utilized to disrupt the standard operation of hardware systems, often to bypass security measures [51]. These attacks can be broadly classified based on the approach employed to introduce faults. They encompass various types, including voltage glitching, clock glitching, electromagnetic fault injection (EMFI), laser fault injection, temperature variation, power cycling, and many more. These attacks typically necessitate advanced technical skills and specialized equipment. Fault injection kits for hobbyists are also known to exist [52].

Generally, these attacks aim to evaluate the resilience of circuits [53]. Nonetheless, they are also aimed at devices storing valuable data or having bypassable security mechanisms, such as smartphones, smart cards, secure tokens, gaming consoles, and various embedded systems.

1.3.7 Side-Channel Attacks

Regarding attacks at the end-user stage, side-channel attacks are a significant concern [54] because they relate to privacy and data protection. Side-channel attacks take advantage of leaked information in the form of current, voltage, timing, acoustic, or electromagnetic emissions. Side-channel attacks typically target cryptographic ICs but also can reveal valuable information for other design implementations [55]. Differential power analysis (DPA) is a side-channel attack that has successfully broken cryptographic implementations [56]. Power samples from the IC under attack are collected for a broad range of plaintext inputs in a DPA attack. Once the samples are gathered, they are subjected to statistical analysis to extract the key. The attack does not aim to break the cryptographic algorithm itself; instead, it targets the implementation and looks for vulnerabilities that allow the extraction of the key bits [57].

It is important to note that many of the aforementioned threats can be effectively mitigated if the design stages are carried out with security measures. However, security threats persist due to the need to share design layout with untrusted foundry and the inevitable aspect that the end-user will have access to the IC. Nevertheless, researchers have developed numerous techniques as countermeasures to combat security threats in the globalized IC supply chain, and the field continues to evolve as researchers strive to introduce novel, practical, and resilient approaches [58].

1.4 Countermeasure Techniques

Countermeasure techniques aimed at enhancing IC security encompass various approaches, typically addressing IP piracy and IC overbuilding concerns. Some techniques also provide indirect protection against hardware trojans. Examples of countermeasure classes include watermarking, fingerprinting, camouflaging, hybrid manufacturing, metering, obfuscation, and ReBO techniques. The following subsection describes these techniques.

1.4.1 Watermarking and Fingerprinting

Watermarking involves embedding a designer's unique signature, such as a secret, into the IC to establish ownership or detect unauthorized modifications [59–61]. On the other hand, fingerprinting incorporates both the designer's and the end-user's signatures to trace instances of piracy [62, 63]. These passive techniques aid in identifying IP piracy but do not actively prevent it. They can be integrated at the logic and physical synthesis stages [62].

1.4.2 Camouflaging

Camouflaging aims to impede RE attempts by replacing specific gates in the design with camouflaged equivalents. When viewed from the top, these camouflaged gates would closely resemble their non-camouflaged counterparts and can implement one of many functions, thus confusing an adversary.

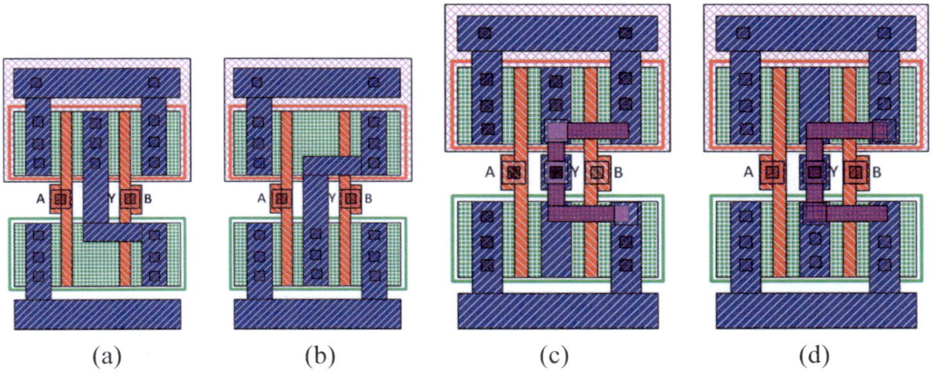

Fig. 1.8 2-input NAND and NOR gates: comparison of layouts for typical and camouflaged 2-input NAND and NOR gates [69]

Camouflaging techniques employ dummy contacts, filler cells, or diffusion-programmable standard cells to achieve their purpose [64]. Upon examination using an optical microscope or a scanning electron microscope (SEM), the actual function of the concealed gates is initially uncertain. Consequently, an RE attacker encounters the challenge of deciphering the purpose of the concealed gates through further investigation [65–68]. In the following example, Fig. 1.8 illustrates camouflaging using dummy contacts. Each camouflaged cell can carry out either a NAND or NOR function. When comparing the metal layers in the traditional layout of 2-input NAND and NOR gates, it is clear that they look distinct from the top, making it easy to identify them through visual inspection. However, in the camouflaged layout of 2-input NAND and NOR gates, the metal layers appear identical, and as a result, the two gates cannot be differentiated from a top view.

1.4.3 Split Manufacturing

Split manufacturing is a sophisticated approach to semiconductor fabrication, where the manufacturing process is intentionally divided and distributed across multiple geographical locations and/or different organizations [70]. The process of split manufacturing is a hybrid IC fabrication approach that involves separating the front-end-of-the-line (FEOL) and the back-end-of-the-line (BEOL) metal layers during the design stage of an IC. The FEOL processing involves the initial steps of semiconductor fabrication, including the layers related to transistor formation [71–73]. The BEOL processing encompasses later stages such as interconnect formation with most of the metal layers. By splitting these stages, companies can send the FEOL view of their designs to one foundry and the BEOL to another. Its primary purpose is to prevent piracy, mainly when dealing with an untrusted foundry. However, it does not offer protection against end-user RE [71].

1.4.4 Metering

Metering techniques are employed to combat the problem of piracy at the user level. These techniques involve assigning a distinct identifier to each IC. Passive metering techniques are used to identify instances of piracy. In contrast, active metering techniques enable the owner of the IC to track and monitor its behavior while it is being used in the field [74].

1.4.5 Logic Locking

LL is generally implemented at the gate-level after the logic synthesis stage by introducing additional logic into a circuit and securing it with key bits. Key bits are stored in a tamper-proof memory and incorporated into the locked circuit with the keys becoming additional inputs. However, selecting the gates to be locked is challenging because not all the locations help to provide better security levels. Figure 1.9 illustrates an example of LL, whereas Fig. 1.9a shows the original netlist.

The additional logic can make use of different combinational cells, such as Mux, AND, OR, XOR, and XNOR gates [75]. The locked circuit functions correctly and generates the expected output only when the correct key value is applied. Otherwise, the circuit's output differs from that of the original design. The locked version of the circuit is demonstrated in Fig. 1.9b and is characterized by three additional XOR/XNOR key gates. Moreover, LL may increase PPA depending upon the technique used. Strategies have been developed to select gates for locking based on maximum security per overhead unit [76].

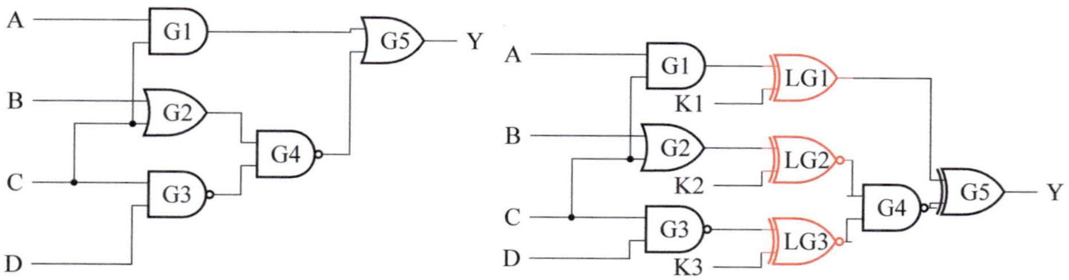

(a) An example circuit that consists of three inputs.

(b) Three XOR/XNOR gates should use the key value of 100.

Fig. 1.9 Logic locking using XOR/XNOR gates

1.4.6 Reconfigurable-Based Obfuscation

ReBO techniques have emerged as highly promising methods that effectively protect against various security threats. Figure 1.10 illustrates the general concept of ReBO with three different devicesbeing constrasted: ASIC design, LUT-based obfuscated design, and a redacted design with embedded FPGA (eFPGA). They represent distinct IC design approaches, with and without obfuscation, each with unique characteristics.

ASICs are specialized chips tailored for specific applications, as depicted in Fig. 1.10. They are meticulously designed to optimize PPA, rendering them highly efficient for their intended functions. ASICs deliver best-in-class performance due to their customized design, eliminating extra components and prioritizing speed and/or power. Their inherent specificity permits minimal power consumption, a critical attribute for battery-powered devices, and their compact design results in minimal chip size, which is advantageous for small-scale electronic devices. However, developing ASICs is costly, involves substantial non-recurring engineering (NRE) expenses and significant investment in design and manufacturing. The development process is protracted, potentially impeding the quick introduction of new products into a market. Once fabricated, ASICs cannot be easily modified to accommodate changes or new features, which limits their adaptability to evolving requirements.

The concept of ReBO is similar to that of an FPGA, where the design is not present until a bitstream is loaded [78]. This approach consists of a reconfigurable part within the circuit, offering robust security measures against untrusted fabrication. A crucial and relatively small part of the circuit remains hidden, taking advantage of its reconfigurable nature. The design is non-functional until the appropriate bitstream is programmed [79]. ReBO techniques can be classified into fine-grain and coarse-grain implementations. For example, LUTs serve fine-grain elements in the center of Fig. 1.10, whereas the right panel ensures the coarse-grain implementation with eFPGA.

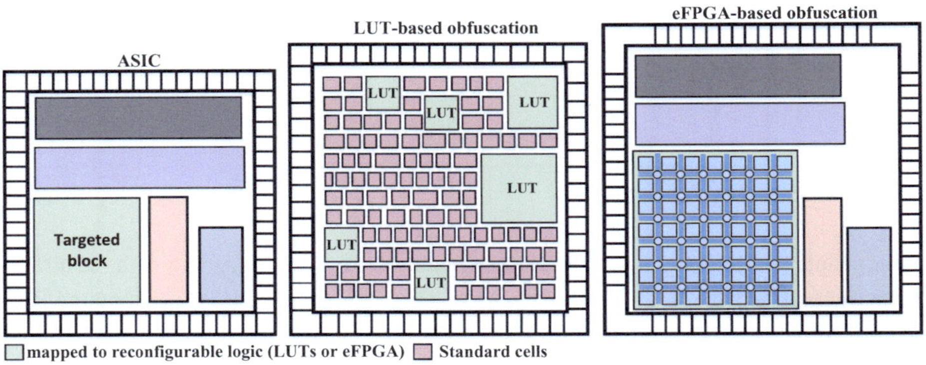

Fig. 1.10 Interpretation of fine-grain and coarse-grain ReBO techniques [77]

A well-known example of fine-grain ReBO implementation is a LUT-based obfuscation that conceals the proper functionality of a circuit using programmable LUTs, as illustrated in the center of Fig. 1.10. From an initial ASIC design, a specific block undergoes conversion into reconfigurable logic for obfuscation. By obfuscating the logic within LUTs, adversaries find it challenging to understand or replicate the design, thus protecting the IP. LUTs can be programmed to complete the design functionality after deployment. Despite these benefits, using LUTs can introduce delays and impact the overall performance of the circuit significantly. Creating and verifying obfuscated circuits is complex and time-consuming, requiring advanced design skills and tools. Additionally, LUT-based designs may consume more power than their non-obfuscated counterparts due to the additional logic that is programmable. Implementing LUT-based obfuscation requires a custom tool and further decisions on how to store the bitstream.

eFPGA redaction is an excellent example of a coarse-grain approach involving eFPGA blocks within an ASIC design that allows parts of the design to remain programmable. eFPGA redaction can protect sensitive parts of the design by keeping them programmable and less transparent, which makes reverse engineering more difficult. Integrating eFPGAs within an ASIC can also be more cost-effective than developing a new ASIC for every change, reducing overall expenses. However, the inclusion of FPGA blocks can result in some performance degradation compared to a fully custom ASIC due to the inherent overhead of programmable logic. eFPGAs typically consume more power than dedicated logic circuits, which can be a drawback for power-sensitive applications. Additionally, integrating and managing ASIC and FPGA components within a single chip adds to the design complexity, potentially making the physical synthesis more challenging to manage with additional floorplan and powerplan requirements.

Both techniques offer effective obfuscation, although they differ significantly in their CAD flow and implementation methodology. Figure 1.10 displays a scaled view, where eFPGA redaction provides the highest overhead among the three approaches. Meanwhile, LUT-based obfuscation falls between ASIC and eFPGA redaction. Both obfuscation techniques require attention in the IC design flow. The choice among these methods depends on the specific requirements of the application, including performance goals and security needs.

1.5 Takeaway Notes

In this chapter, various supply chain related threats were introduced, along with the different countermeasure techniques to overcome them. The chapter closes by explaining the ReBO method, which is the centerpiece of this book. In Chap. 2, we will present insights into the current scenario of the semiconductor industry and the evolution of reconfigurable logic.

References

1. I. Bojanova, "The digital revolution: What's on the horizon?," *IT Professional*, vol. 16, no. 1, pp. 8–12, 2014.
2. PHYSEC, "Digitalization of critical infrastructure," last accessed on Sep 11, 2023. Available at: https://www.physec.de/en/manufacturer/industry/digitalization-of-critical-infrastructure/.
3. Y.-Q. Lv, Q. Zhou, Y.-C. Cai, and G. Qu, "Trusted integrated circuits: The problem and challenges," *Journal of Computer Science and Technology*, vol. 29, pp. 918–928, Sep 2014.
4. IEEE Digital Reality, "The impacts that digital transformation has on society," last accessed on Feb 02, 2023. Available at: https://digitalreality.ieee.org/publications/impacts-of-digital-transformation.
5. International Roadmap for Devices and Systems, "Semiconductors and artificial intelligence," last accessed on Mar 02, 2023. Available at: https://irds.ieee.org/topics/semiconductors-and-artificial-intelligence.
6. Electronicshub, "Integrated circuits-types , uses," last accessed on Mar 30, 2023. Available at: https://www.electronicshub.org/integrated-circuits-types-uses/.
7. NobelPrize.org, "Nobel lecture: Miniaturization of electronic circuits–the past and the future," last accessed on Apr 30, 2023. Available at: https://www.nobelprize.org/prizes/physics/2000/kilby/lecture/.
8. J. Bardeen and W. Brattain, "The transistor, a semiconductor triode," *Proceedings of the IEEE*, vol. 86, no. 1, pp. 29–30, 1998.
9. Industrial Alchemy, "Texas instruments SN514 integrated circuit," last accessed on Sep 20, 2023. Available at: https://www.industrialalchemy.org/articleview.php?item=741.
10. HC (History Computer), "Integrated circuit (IC) explained — everything you need to know," last accessed on May 30, 2023. Available at: https://history-computer.com/integrated-circuit/.
11. T. Perry, "Intel's secret is out," *IEEE Spectrum*, vol. 26, no. 4, pp. 22–28, 1989.
12. The Silicon Engine, "The timeline of semiconductors in computers," last accessed on Sep 2, 2023. Available at: https://www.computerhistory.org/siliconengine/.
13. P. Chaturvedi, "Wafer scale integration: a review," *Microelectronics Journal*, vol. 19, no. 2, pp. 4–35, 1988.
14. O. Doug, "Lessons from history: The 1980s semiconductor cycle(s)," last accessed on May 08, 2022. Available at: https://www.fabricatedknowledge.com/p/history-lesson-the-1980s-semiconductor.
15. K. William W. and P. Louis W., "Crisis and adaptation in east asian innovation systems: The case of the semiconductor industry in taiwan and south korea," *Business & Politics*, vol. 2, no. 3, pp. 327–352, 2000.
16. L. Alberto, "What is a fabless chip company?," last accessed on Mar 29, 2023. Available at: https://miscircuitos.com/fabless/.
17. S. Stephen, "Moore's law: The rule that really matters in tech," last accessed on Mar 22, 2023. Available at: https://www.cnet.com/science/moores-law-the-rule-that-really-matters-in-tech/.
18. F. Nate, "Who's going to pay for american-made semiconductors?," last accessed on Mar 20, 2023. Available at: https://builtin.com/hardware/american-made-semiconductor-costs.
19. C. A. Johan, G. Dieter, H. Guido, H. Denis, and H. Arndt, "How does the semiconductor industry landscape look today?," last accessed on Mar 20, 2023. Available at: https://www.kearney.com/industry/technology/article/-/insights/how-does-the-semiconductor-industry-landscape-look-today.
20. A. Majeed, "The truth about smic's 7-nm chip fabrication ordeal," last accessed on Mar 26, 2023. Available at: https://www.edn.com/the-truth-about-smics-7-nm-chip-fabrication-ordeal/.

21. S. Matthew and R. William Alan, "Contextualizing the national security concerns over china's domestically produced high-end chip." https://www.csis.org/analysis/contextualizing-national-security-concerns-over-chinas-domestically-produced-high-end-chip, 2023. Accessed: June 17, 2024.
22. T. Charlotte, "Us government still investigating alleged chip sanctions breach by smic." https://www.datacenterdynamics.com/en/news/us-government-still-investigating-alleged-chip-sanctions-breach-by-smic/, 2024. Accessed: June 17, 2024.
23. IC Insights, "U.S. chip suppliers continue to dominate R&D spending," last accessed on Mar 27, 2023. Available at: https://www.eetasia.com/u-s-chip-suppliers-continue-to-dominate-rd-spending/.
24. D. Paula, "Process complexity means exponentially increasing data volumes and analysis challenges," last accessed on May 12, 2023. Available at: http://bit.ly/3t6IBwY.
25. D. Shannon, "Shortage to surplus cycle hits semi but one segment escapes," last accessed on May 18, 2023. Available at: https://www.semiconductor-digest.com/shortage-to-surplus-cycle-hits-semi-but-one-segment-escapes/.
26. S. Steffen, "Using AI to pattern sub-10nm ICs," last accessed on Sep 1, 2023. Available at: https://www.ednasia.com/using-ai-to-pattern-sub-10nm-ics/.
27. Y. Badr, A. Torres, and P. Gupta, "Mask assignment and DSA grouping for DSA-MP hybrid lithography for sub-7 nm contact/via holes," *IEEE Transactions on Computer-Aided Design of Integrated Circuits and Systems*, vol. 36, no. 6, pp. 913–926, 2017.
28. S. Ed, "Design rule complexity rising," last accessed on May 30, 2023. Available at: https://semiengineering.com/design-rule-complexity-rising/.
29. International Roadmap for Devices and Systems, "High-end performance packaging 2022 – focus on 2.5d/3d integration," last accessed on Mar 03, 2023. Available at: https://www.yolegroup.com/product/report/high-end-performance-packaging-2022--focus-on-25d3d-integration/.
30. F. Alan, "TSMC's new 2nm chip production fab will cost it how much?," last accessed on Jun 20, 2022. Available at: https://www.phonearena.com/news/tsmc-to-spend-fortune-on-2nm-production-fab_id140626.
31. M. Yasin, J. J. Rajendran, and O. Sinanoglu, "The need for logic locking," in *Trustworthy Hardware Design: Combinational Logic Locking Techniques*, pp. 1–16, Cham: Springer International Publishing, 2020.
32. Intel, "Responsible, resilient, and diverse supply chain." https://www.intel.com/content/www/us/en/corporate-responsibility/supply-chain.html, 2023. Accessed: April 12, 2024.
33. MacRumors, "Apple books nearly 90% of tsmc's 3nm production capacity for this year," last accessed on Jun 24, 2022. Available at: https://www.macrumors.com/2023/05/15/apple-tsmc-3nm-production-capacity/.
34. M. Rostami, F. Koushanfar, and R. Karri, "A primer on hardware security: Models, methods, and metrics," *Proceedings of the IEEE*, vol. 102, no. 8, pp. 1283–1295, 2014.
35. S. Agam, "Europe, US warn of fake-chip danger to national security, critical systems," last accessed on Feb 02, 2023. Available at: https://www.theregister.com/2022/03/18/eu_us_counterfeit_chips/.
36. A. Matthew, "Supply chain threats against integrated circuits," last accessed on Jan 02, 2023. Available at: https://www.intel.com/content/dam/www/public/us/en/documents/white-papers/supply-chain-threats-v1.pdf.
37. B. Lippmann, A.-C. Bette, M. Ludwig, J. Mutter, J. Baehr, A. Hepp, H. Gieser, N. Kovač, T. Zweifel, M. Rasche, and O. Kellermann, "Physical and functional reverse engineering challenges for advanced semiconductor solutions," in *Proceedings of the 2022 Conference & Exhibition on Design, Automation & Test in Europe*, DATE '22, (Leuven, BEL), p. 796–801, European Design and Automation Association, 2022.

38. R. S. Rajarathnam, Y. Lin, Y. Jin, and D. Z. Pan, "ReGDS: A reverse engineering framework from GDSII to gate-level netlist," in *2020 IEEE International Symposium on Hardware Oriented Security and Trust (HOST)*, pp. 154–163, 2020.
39. P. Subramanyan, N. Tsiskaridze, K. Pasricha, D. Reisman, A. Susnea, and S. Malik, "Reverse engineering digital circuits using functional analysis," in *Proceedings of the Conference on Design, Automation and Test in Europe*, DATE '13, (San Jose, CA, USA), p. 1277–1280, EDA Consortium, 2013.
40. N. Albartus, M. Hoffmann, S. Temme, L. Azriel, and C. Paar, "DANA universal dataflow analysis for gate-level netlist reverse engineering," *IACR Transactions on Cryptographic Hardware and Embedded Systems*, vol. 2020, no. 4, pp. 309–336, 2020.
41. R. Kibria, M. Sazadur Rahman, F. Farahmandi, and M. Tehranipoor, "RTL-FSMx: Fast and accurate finite state machine extraction at the RTL for security applications," in *2022 IEEE International Test Conference (ITC)*, pp. 165–174, 2022.
42. F. Koushanfar, "Integrated circuits metering for piracy protection and digital rights management: An overview," in *Proceedings of the 21st Edition of the Great Lakes Symposium on Great Lakes Symposium on VLSI*, GLSVLSI '11, (New York, NY, USA), p. 449–454, Association for Computing Machinery, 2011.
43. M. Xue, C. Gu, W. Liu, S. Yu, and M. O'Neill, "Ten years of hardware trojans: a survey from the attacker's perspective," *IET Computers & Digital Techniques*, vol. 14, no. 6, pp. 231–246, 2020.
44. Y. Jin, N. Kupp, and Y. Makris, "Experiences in hardware trojan design and implementation," in *2009 IEEE International Workshop on Hardware-Oriented Security and Trust*, pp. 50–57, 2009.
45. R. Karri, J. Rajendran, K. Rosenfeld, and M. Tehranipoor, "Trustworthy hardware: Identifying and classifying hardware trojans," *Computer*, vol. 43, no. 10, pp. 39–46, 2010.
46. C. Sturton, M. Hicks, D. Wagner, and S. T. King, "Defeating uci: Building stealthy and malicious hardware," in *2011 IEEE Symposium on Security and Privacy*, pp. 64–77, 2011.
47. T. D. Perez and S. Pagliarini, "Hardware trojan insertion in finalized layouts: From methodology to a silicon demonstration," *IEEE Transactions on Computer-Aided Design of Integrated Circuits and Systems*, vol. 42, no. 7, pp. 2094–2107, 2023.
48. SECLORE, "Intellectual property (IP) theft in the semiconductor industry: Innovation at risk," last accessed on Sep 20, 2023. Available at: https://www.seclore.com/wp-content/uploads/2023/08/Seclore-Intellectual-Property-IP-Theft-in-Semiconductor-Industry.pdf.
49. U. Guin, K. Huang, D. DiMase, J. M. Carulli, M. Tehranipoor, and Y. Makris, "Counterfeit integrated circuits: A rising threat in the global semiconductor supply chain," *Proceedings of the IEEE*, vol. 102, no. 8, pp. 1207–1228, 2014.
50. EUIPO-ITU, "EUIPO-ITU report: The economic cost of IPR infringement in the smartphones sector," last accessed on Mar 02, 2023. Available at: https://www.itu.int/en/ITU-D/Regulatory-Market/Pages/Counterfeiting/SmartphonesStudy.aspx.
51. Risecure, "What is fault injection?." https://www.riscure.com/fault-injection/, 2024. Accessed: June 18, 2024.
52. NewAE Technology Inc., "Chipshouter kit." https://www.newae.com/products/nae-cw520, 2024. Accessed: August 16, 2024.
53. Z. U. Abideen and M. Rashid, "Efic-me: A fast emulation based fault injection control and monitoring enhancement," *IEEE Access*, vol. 8, pp. 207705–207716, 2020.
54. F. Koeune and F.-X. Standaert, "A tutorial on physical security and side-channel attacks," in *Foundations of Security Analysis and Design III: FOSAD 2004/2005 Tutorial Lectures* (A. Aldini, R. Gorrieri, and F. Martinelli, eds.), pp. 78–108, Berlin, Heidelberg: Springer Berlin Heidelberg, 2005.

55. B. Yang, K. Wu, and R. Karri, "Scan based side channel attack on dedicated hardware implementations of data encryption standard," in *2004 International Conferce on Test*, pp. 339–344, 2004.
56. P. Kocher, J. Jaffe, and B. Jun, "Differential power analysis," in *Advances in Cryptology — CRYPTO' 99* (M. Wiener, ed.), (Berlin, Heidelberg), pp. 388–397, Springer Berlin Heidelberg, 1999.
57. Rambus Press, "Side-channel attacks explained: everything you need to know," last accessed on Aug 27, 2023. Available at: https://www.rambus.com/blogs/side-channel-attacks/#what.
58. S. Engels, M. Hoffmann, and C. Paar, "A critical view on the real-world security of logic locking," *Journal of Cryptographic Engineering*, vol. 12, pp. 229–244, Sep 2022.
59. A. B. Kahng, J. Lach, W. H. Mangione-Smith, S. Mantik, I. L. Markov, M. Potkonjak, P. Tucker, H. Wang, and G. Wolfe, "Watermarking techniques for intellectual property protection," in *Proceedings of the 35th Annual Design Automation Conference*, DAC '98, (New York, NY, USA), p. 776–781, Association for Computing Machinery, 1998.
60. M. Khan and S. Tragoudas, "Rewiring for watermarking digital circuit netlists," *IEEE Transactions on Computer-Aided Design of Integrated Circuits and Systems*, vol. 24, no. 7, pp. 1132–1137, 2005.
61. M. Lewandowski, R. Meana, M. Morrison, and S. Katkoori, "A novel method for watermarking sequential circuits," in *2012 IEEE International Symposium on Hardware-Oriented Security and Trust*, pp. 21–24, 2012.
62. X. Chen, G. Qu, and A. Cui, "Practical IP watermarking and fingerprinting methods for ASIC designs," in *2017 IEEE International Symposium on Circuits and Systems (ISCAS)*, pp. 1–4, 2017.
63. H. Huang, A. Boyer, and S. B. Dhia, "The detection of counterfeit integrated circuit by the use of electromagnetic fingerprint," in *2014 International Symposium on Electromagnetic Compatibility*, pp. 1118–1122, 2014.
64. M. Yasin, B. Mazumdar, O. Sinanoglu, and J. Rajendran, "Removal attacks on logic locking and camouflaging techniques," *IEEE Transactions on Emerging Topics in Computing*, vol. 8, no. 2, pp. 517–532, 2020.
65. "Method and apparatus for camouflaging a standard cell based integrated circuit with micro circuits and post processing," 2013. last accessed on May 30, 2024.
66. R. P. Cocchi, J. P. Baukus, L. W. Chow, and B. J. Wang, "Circuit camouflage integration for hardware IP protection," in *2014 51st ACM/EDAC/IEEE Design Automation Conference (DAC)*, pp. 1–5, 2014.
67. M. Li, K. Shamsi, T. Meade, Z. Zhao, B. Yu, Y. Jin, and D. Z. Pan, "Provably secure camouflaging strategy for IC protection," *IEEE Transactions on Computer-Aided Design of Integrated Circuits and Systems*, vol. 38, no. 8, pp. 1399–1412, 2019.
68. J. Rajendran, M. Sam, O. Sinanoglu, and R. Karri, "Security analysis of integrated circuit camouflaging," in *Proceedings of the 2013 ACM SIGSAC Conference on Computer & Communications Security*, CCS '13, p. 709–720, Association for Computing Machinery, 2013.
69. J. Rajendran, O. Sinanoglu, and R. Karri, "Vlsi testing based security metric for ic camouflaging," in *2013 IEEE International Test Conference (ITC)*, pp. 1–4, 2013.
70. Y. Wang, P. Chen, J. Hu, and J. J. Rajendran, "The cat and mouse in split manufacturing," in *Proceedings of the 53rd Annual Design Automation Conference*, DAC '16, (New York, NY, USA), Association for Computing Machinery, 2016.
71. T. D. Perez and S. Pagliarini, "A survey on split manufacturing: Attacks, defenses, and challenges," *IEEE Access*, vol. 8, pp. 184013–184035, 2020.
72. J. Rajendran, O. Sinanoglu, and R. Karri, "Is split manufacturing secure?," in *2013 Design, Automation Test in Europe Conference Exhibition (DATE)*, pp. 1259–1264, 2013.

73. A. Sengupta, M. Nabeel, J. Knechtel, and O. Sinanoglu, "A new paradigm in split manufacturing: Lock the feol, unlock at the beol," in *2019 Design, Automation & Test in Europe Conference & Exhibition (DATE)*, pp. 414–419, 2019.
74. F. Koushanfar, "Hardware metering: A survey," in *Introduction to Hardware Security and Trust* (M. Tehranipoor and C. Wang, eds.), pp. 103–122, New York, NY: Springer New York, 2012.
75. M. Yasin, J. Rajendran, and O. Sinanoglu, "Trustworthy hardware design: Combinational logic locking techniques," Springer, Cham, 2019.
76. K. Zamiri Azar, H. Mardani Kamali, H. Homayoun, and A. Sasan, "Threats on logic locking: A decade later," in *GLSVLSI '19: Proceedings of the 2019 on Great Lakes Symposium on VLSI*, p. 471–476, 2019.
77. Z. U. Abideen, S. Gokulanathan, M. J. Aljafar, and S. Pagliarini, "An overview of FPGA-inspired obfuscation techniques," *ACM Comput. Surv.*, jul 2024. Just Accepted.
78. A. Baumgarten, A. Tyagi, and J. Zambreno, "Preventing IC piracy using reconfigurable logic barriers," *IEEE Design Test of Computers*, vol. 27, no. 1, pp. 66–75, 2010.
79. B. Liu and B. Wang, "Embedded reconfigurable logic for ASIC design obfuscation against supply chain attacks," in *2014 Design, Automation Test in Europe Conference Exhibition (DATE)*, pp. 1–6, 2014.

Emergence of Reconfigurable Logic

2

2.1 Introduction to FPGA, ASIC and Structured ASIC

System architects have a variety of platforms to consider for hardware design. These options include FPGAs, structured ASICs, and ASICs, all of which can be customized to a given application. To meet their specific needs for flexibility, performance, power efficiency, and overall cost of ownership, as well as to align with time-to-market requirements, architects must carefully select the hardware type that best fits their unique circumstances.

2.1.1 FPGA Concept

An FPGA is a semi-custom integrated circuit, where programmable units are pre-defined by the FPGA vendors. Once the device is delivered to the end users, it can be customized to fully implement the desired functionality. FPGAs offer engineers and developers a highly flexible hardware platform that can be adapted to meet various application requirements.

The flexibility of FPGAs can be analogous to that of a digital chameleon. Just as chameleons can change their colors and patterns to blend into different environments, FPGAs can adapt their hardware configurations to meet diverse and evolving application needs. This comparison highlights FPGAs' dynamic ability to transform and optimize themselves, similar to how a chameleon adjusts its appearance to blend with its surroundings. This makes FPGAs highly valuable in industries where rapid customization, performance optimization, and scalability are essential.

Z. U. Abideen and S. Pagliarini, *Reconfigurable Obfuscation Techniques for the IC Supply Chain*, Synthesis Lectures on Digital Circuits & Systems,
https://doi.org/10.1007/978-3-031-77509-3_2

2.1.2 ASIC Concept

ASICs are purpose-built and mass-produced for a specific function, offering higher performance and efficiency than FPGA. At very high volumes, ASICs present a lower price per unit than FPGAs as the initially high NRE costs are diluted. Unlike FPGAs, ASICs cannot be reprogrammed once manufactured, necessitating a significant NRE re-investment to make any design changes.

ASICs typically contain a mix of custom blocks and cell-based blocks. Custom blocks, as the name implies, are drawn by hand by layout engineers. Examples of such blocks include converters, clock generators, and memories. Cell-based blocks make use of digital standard cells that are laid out in rows. All cells have the same height for the sake of compatibility and automation. A digital ASIC designer will resort to place and route tools to implement his/her design during physical synthesis.

2.1.3 Structured ASIC Concept

Large NRE costs and high minimum order quantities have always been a challenge for ASICs. This is especially true for ASICs with advanced technologies. FPGAs provide some solutions, but their performance and cost limitations have led to the development of a category of devices called structured ASICs.

Unlike standard cell ASICs, which require a new mask set for each design, structured ASICs allow customization only on the top layers, reducing the need for custom masks. These types of ASICs also reduce the number of programmable interconnect layers and incorporate embedded IPs such as RAM and timing generators, as well as pre-designed features like embedded clocks and test structures, to minimize design cycle time. A structured ASIC connects array cells to create the desired circuit. Most structured ASIC architectures use an array cell structure repeated multiple times on the die to form the structured fabric.

There are three main types of structured ASIC architectures in use [1]. The fine-grained architecture incorporates basic array cells similar in structure to a classic four-transistor gate-array core cell, with higher-level functions such as I/O cells and test structures fully formed and embedded within the fabric. The medium-grained architecture features an array cell comprised of various combinations of simple logic functions and RAM, complex enough to implement a simple FF in a single cell, with pre-diffused blocks of IP surrounding the array cells. The coarse-grained architecture uses a RAM-based LUT and D-FF to define the logic functionality, with array cells grouped into blocks and arrayed across the chip and programmable interconnect layers used to connect the cells as desired.

2.2 Introduction to FPGA Architecture

The architecture of an FPGA comprises several key components that work together to provide its reprogrammable capabilities. These components include configurable logic blocks (CLBs),[1] interconnects, input/output (I/O) blocks, block RAM, digital signal processing (DSP) blocks, and clock management resources [2]. The interconnect network is crucial in linking different building blocks and can be customized to support various routing configurations based on the specific use case. The diagram depicted in Fig. 2.1 illustrates the configuration of a conventional island-style FPGA. The input/output blocks (IOBs) interface the FPGA and the external world. Among all of these components, CLB plays a pivotal role in enabling the programmable nature within the FPGA architecture, allowing designers to configure them to implement various logic functions[2] such as AND, OR, XOR gates, flip-flops, and more complex functions.

2.2.1 Internal Architecture of a CLB

Figure 2.2 shows the internal structure of a commercial CLB, like the ones in Xilinx devices (now AMD). Each CLB includes two FFs and two independent 4-input function generators, offering designers flexibility as most logic functions require fewer than four inputs. This allows design software to handle each generator independently, optimizing cell usage. Thirteen inputs and four outputs per CLB provide access to flip-flops and function generators, doubling the XC3000 families' resources. Each of the two function generators (F1 – F4 and G1 – G4) receives four independent inputs and outputs of LUTs A and B, capable of implementing any Boolean function of their inputs using memory LUTs with consistent

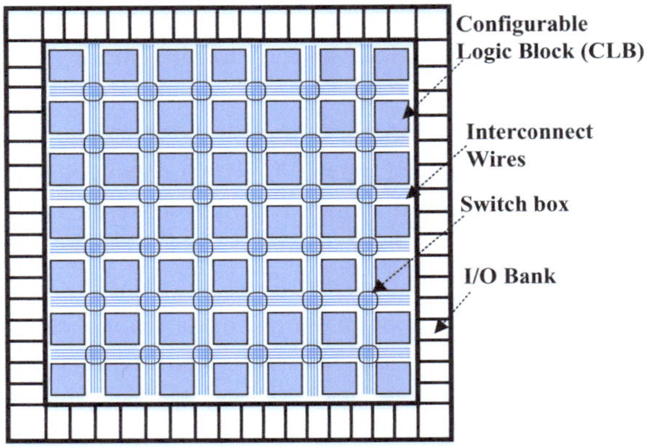

Fig. 2.1 The traditional island-style architecture of FPGA [3]

[1] There are slightly different terminologies from vendor to vendor, but the concepts are the same.

[2] Logic functions are implemented on LUTs A, B and (A, B).

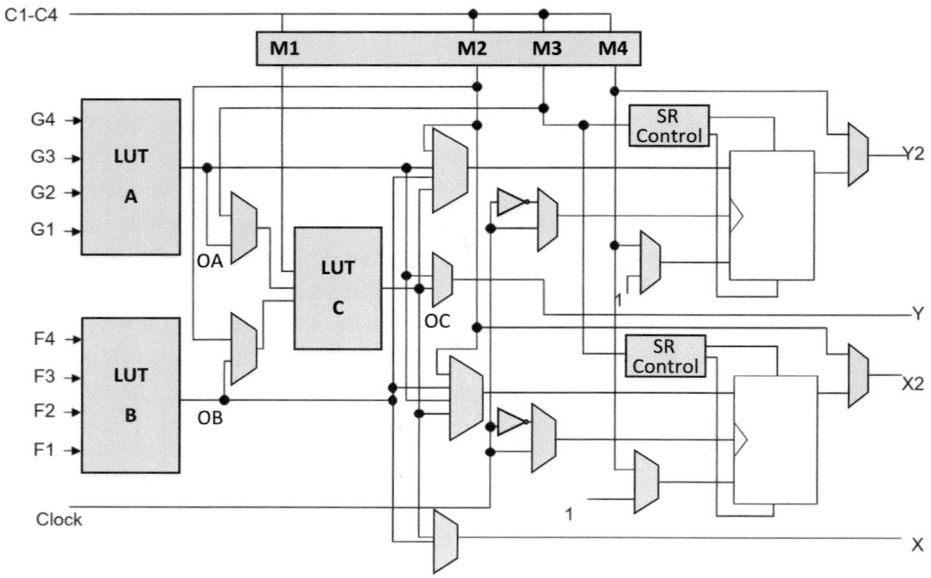

Fig. 2.2 Standard CLB architecture

propagation delay. A third LUT C receives two inputs from Muxes and an external input (M1). A shared input (SR) can be configured to act as an asynchronous set or reset signal for each of the two registers independently. It can also be disabled for either flip-flop. Additionally, a dedicated global set/reset line clears each register during power-up, reconfiguration, or when a specific Reset net is activated.

Outputs from the generators exit through two ports: A or B can connect to X, and C or A to Y. This enables a CLB to implement two independent functions with up to four variables each or single functions with up to nine variables. Implementing wide functions within a block reduces the number required and delays in the signal path, increasing density and speed. The CLB's storage comprises edge-triggered D-FFs with a common clock, and clock enable inputs. A third input (M3) can independently program asynchronous set/reset signals for each FF and be disabled. A global set/reset line manages power-up and reconfiguration without competing for routing resources, connecting to any package pin as a global reset input.

Each flip-flop operates on rising or falling clock edges, programmable and driven by A, B, (A, B), or "direct in" (M2). The FFs drive outputs X2 and Y2. Furthermore, CLB function generators A and B feature dedicated arithmetic logic for rapid carry and borrow signal generation, boosting efficiency in operations like adders, subtracters, accumulators, comparators, and counters. Multiplexers in the CLB map four control inputs (C1-C4) to internal signals (M1, M2, M3, M4) flexibly, aiding placement and routing by treating function generators and FFs independently for high packing density. This architecture's symmetry optimizes application layout, with inputs, outputs, and functions freely repositioned to avoid congestion during placement and routing.

2.2.2 Boolean Implementation Using LUT

The previous section provided an overview of CLB in FPGAs. CLBs can consist of either a single basic logic element (BLE) or a cluster of locally interconnected BLEs. Modern FPGAs typically contain CLBs with multiple BLEs. A basic BLE is comprised of a LUT and a FF. The LUT is a fundamental component of the CLB that is responsible for implementing boolean functions. A k-input LUT contains 2^k configuration bits and has the capability to implement any k-input boolean function. A 4-input LUT utilizes 16 static random access memory (SRAM) bits to realize any 4-input boolean function. The LUT's output is optionally connected to a FF, with a multiplexer (Mux) selecting the output to be from either the FF or the LUT. Figure 2.3 illustrates a simple BLE with a 4-input LUT and a D-FF.

Moreover, in the context of the CLB as depicted in Fig. 2.2, a maximum of 4-input boolean functions can be implemented. A 4-input LUT has four binary inputs (A, B, C, D) and one binary output (F), encompassing 16 memory cells to cover the entire range of possible input combinations. Essentially, the LUT uses a truth table to specify the output for each of these 16 input combinations.

The operation of a 4-input LUT involves applying the four input lines (A, B, C, D) to select one of the 16 memory cells. The combination of input values forms an address pointing to a specific LUT memory cell. The value stored in the addressed memory cell is the output

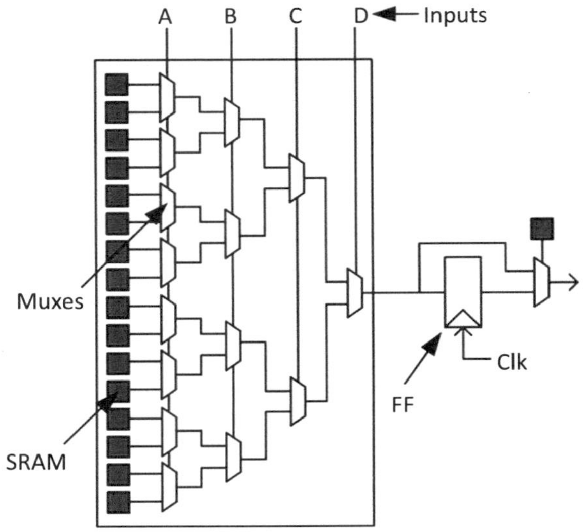

Fig. 2.3 Internal structure of BLE [4]

of the LUT for that particular combination of inputs. For example, if the inputs are (A=1, B=0, C=1, D=1), the address is 1011 in binary, corresponding to the 11th memory cell. The value stored in this cell is the output of the LUT. To program a 4-input LUT, you need to define the output for each of the 16 possible input combinations. This is usually done using a truth table. Table 2.1 represents the boolean functions that the 4-input LUT can implement. This truth table is programmed into the LUT so that the corresponding value of F is output for each combination of inputs (A, B, C, D).

Table 2.1 Possible functions implementable by a 4-input LUT in different forms

A	B	C	D	Sum of products	Conjunctive normal form (CNF)
0	0	0	0	0	1
0	0	0	1	’A’B’C D	(A ∨ B ∨ C) ∧ (A ∨ B ∨ ¬D) ∧ (A ∨ ¬C ∨ ¬D) ∧ (¬A ∨ B ∨ ¬D)
0	0	1	0	’A’B C’D	(A ∨ B ∨ ¬C) ∧ (A ∨ ¬C ∨ D) ∧ (A ∨ B ∨ D) ∧ (¬A ∨ B ∨ D)
0	0	1	1	’A’B C D	(A ∨ B ∨ ¬C) ∧ (A ∨ ¬C ∨ ¬D) ∧ (¬A ∨ B ∨ ¬C)
0	1	0	0	’A B’C’D	(A ∨ ¬B ∨ C) ∧ (¬A ∨ ¬B ∨ D) ∧ (¬A ∨ C ∨ D)
0	1	0	1	’A B’C D	(A ∨ ¬B ∨ C) ∧ (¬A ∨ ¬B ∨ ¬D) ∧ (¬A ∨ C ∨ ¬D)
0	1	1	0	’A B C’D	(A ∨ ¬B ∨ ¬C) ∧ (¬A ∨ ¬B ∨ D) ∧ (¬A ∨ ¬C ∨ D)
0	1	1	1	’A B C D	(A ∨ ¬B ∨ ¬C) ∧ (¬A ∨ ¬B ∨ ¬D)
1	0	0	0	A’B’C’D	(¬A ∨ B ∨ C) ∧ (A ∨ B ∨ D) ∧ (A ∨ C ∨ D)
1	0	0	1	A’B’C D	(¬A ∨ B ∨ C) ∧ (A ∨ B ∨ ¬D) ∧ (A ∨ C ∨ ¬D)
1	0	1	0	A’B C’D	(¬A ∨ B ∨ ¬C) ∧ (A ∨ B ∨ D) ∧ (A ∨ ¬C ∨ D)
1	0	1	1	A’B C D	(¬A ∨ B ∨ ¬C) ∧ (A ∨ B ∨ ¬D) ∧ (A ∨ ¬C ∨ ¬D)
1	1	0	0	A B’C’D	(¬A ∨ ¬B ∨ C) ∧ (¬A ∨ C ∨ D)
1	1	0	1	A B’C D	(¬A ∨ ¬B ∨ C) ∧ (¬A ∨ ¬C ∨ D)
1	1	1	0	A B C’D	(¬A ∨ ¬B ∨ ¬C) ∧ (¬A ∨ ¬C ∨ D)
1	1	1	1	A B C D	(¬A ∨ ¬B ∨ ¬C) ∧ (¬A ∨ ¬B ∨ ¬D)

Despite their advantages, LUTs have limitations. LUT in FPGA designs have limited resolution based on the truth table. They can consume significant FPGA resources, leading to potential trade-offs with other elements. While providing fast execution, they introduce propagation delay and compete for resources. Altering the logic function may require reprogramming the LUT, limiting its reusability. Additionally, larger logic functions may require multiple LUTs or more complex configurations, impacting resource utilization and design complexity.

2.2.3 FPGA Routing

As discussed earlier, the computing functionality is enabled through programmable logic blocks interconnected via a programmable routing network. Hundreds of thousands of configurable logic blocks of various types permit the implementation of highly intricate devices on a single chip, facilitating easy configuration. Modern FPGAs are equipped with sufficient capacity to accommodate multiple complex processors on a single device. The programmable routing network facilitates establishing connections among logic blocks and I/O blocks to realize user-defined circuits. The routing interconnects within an FPGA comprise wires and programmable switches that facilitate the establishment of requisite connections, with the configuration of these switches being achieved using programmable technology.

FPGA architectures are often considered as potential candidates for implementing various digital circuits due to their potential for accommodating a wide range of routing demands. This necessitates a highly flexible routing interconnect to cater to the diverse requirements of different circuits. Despite the variations in routing demands across different circuits, certain common characteristics can be leveraged to optimize the design of the routing interconnect in FPGA architectures. For instance, most designs exhibit local connectivity, necessitating an abundance of short wires. However, there also exist distant connections (e.g., busses), which require sparse, long wires. Therefore, careful consideration is essential in the design of routing interconnects for FPGA architectures to ensure both flexibility and efficiency.

The arrangement of routing resources in relation to the layout of logic blocks within the architecture profoundly influences its overall efficiency. This arrangement is referred to as the global routing architecture, while the detailed routing architecture pertains to the details of the switching topology of different switch blocks. Based on the global arrangement of routing resources, FPGA architectures can be classified as either hierarchical [5] or island-style [6].

2.3 Evolution of FPGA Hardware

Xilinx Corporation (since then acquired by AMD in 2022) invented the first commercial FPGA—the XC2064 device—in 1985. This device is illustrated in Fig. 2.4. The device contains 64 programmable logic units consisting of two 3-input LUTs and a FF, enabling

Fig. 2.4 Worlds first FPGA device [8]

true "field" programmability for the first time. IEEE listed Xilinx XC2064 as one of the "25 Microchips that Shook the World" in 2009 [7] and inducted it into the "Chip Hall of Fame" in 2017 [8].

Starting from its inception, the FPGA industry has consistently remained at the leading edge of semiconductor technology. Over time, it has developed a distinctive and intricate knowledge ecosystem that encompasses FPGA chip design, FPGA chip design EDA, FPGA application design, FPGA application design EDA, FPGA assembly and testing, as well as FPGA sales and marketing, among others. With more and more heterogeneous units integrated into FPGAs, this knowledge system is expanding at an unprecedented speed. FPGA vendors, especially the market leaders, have led the innovation most of the time due to their strong ability to integrate supply chain resources from upstream to downstream and maintain a diverse and interdisciplinary talent echelon. FPGA hardware evolution can be inspected from several perspectives, which we specify in the following subsections.

2.3.1 Bitstream Storage

Configuration memory is a crucial component of FPGAs as it stores the configuration bit data, often simply called bitstream. FPGAs can be classified into different types based on their configuration memory.

The SRAM type is the most common type of FPGA since its inception. The configuration data needs to be loaded into the on-chip SRAM when the power is turned on. Once the device is turned off, the data in the SRAM is lost (volatile), and the internal logic function of the FPGA also disappears. SRAM-based FPGAs are reusable and cost-effective, but may require the assistance of an external memory device that is not volatile.

On the other hand, the Antifuse type FPGA can only be programmed once by burning the fuses within. While it loses the flexibility of reprogramming, it significantly improves stability. These FPGAs are particularly suitable for applications in harsh environments such as strong electromagnetic radiation. Due to the fixed logic, these devices power up instantaneously and consume less power and space compared to other types of FPGAs.

Table 2.2 Comparison of FPGAs based on different memory types

Memory type	Volatile	Programmability	Latency	Power	Area	Cost
SRAM	Yes	Repeatable	Low	Medium	Large	Low
Antifuse	No	Once	Very low	Very low	Very small	Very high
Flash	No	Repeatable	High	Low	Small	Medium
MRAM	No	Repeatable	Low	High	Small	High
RRAM	No	Repeatable	Low	High	Small	High

Flash memory is a type of non-volatile memory that combines the flexibility of an SRAM memory with the non-volatility of an anti-fuse device. While this technology is costly, it uses a relatively small number of transistors and has low leakage current [9]. These devices are well-suited for aerospace applications [10–12].

In recent years, there has been active research into emerging non-volatile FPGA technology in order to further improve PPA. Two promising non-volatile memory devices that have the potential to replace Flash memory in FPGAs are resistive random access memory (RRAM) [13] and magnetoresistive random access memory (MRAM) [14]. Both RRAM and MRAM technologies offer several advantages over Flash memory. First, they are compatible with the BEOL process, resulting in significantly higher area density. Unlike flash transistors, RRAMs and MRAMs are fabricated between metal layers, eliminating transistor area consumption [15, 16]. Secondly, RRAM and MRAM technologies operate at low read/write voltages and currents, similar to the nominal voltages of logic transistors. This eliminates the need for high-voltage circuitry for memory writing and the overheads that come with it [15–17]. Finally, both RRAM and MRAM technologies offer fast read and write speeds at the nanosecond scale, which can reduce programming time and power consumption in FPGA devices [18, 19]. Currently, FPGAs based on emerging technologies have garnered significant research interest and are seen as the next generation of FPGA technology, as outlined in Table 2.2.

2.3.2 Logic Tile Structure

The range of component resource types is expanding quickly due to advancements in programmable technology. Modern FPGAs have transformed into hybrid devices with various tile architectures.

Most FPGA vendors commonly denote the fundamental programmable unit within the FPGA as the CLB/logic array block (LAB). As described earlier, the CLB typically contains BLEs, which comprise LUT/FF, and its structure can vary. For instance, in the case of Xilinx (AMD) FPGAs, the CLB is the primary resource for implementing basic sequential and combinational circuit functions [20]. In the Versal architecture, also from AMD, each CLB contains 4 BLEs, with each BLE housing 8 adaptive 6-input LUTs and 16 FFs [21]. Xilinx CLBs are available in two types: SliceL (Logic) and SliceM (Memory). The latter allows the LUT under the CLB to function as distributed memory by incorporating independent write addresses, write enable and clock signals. This enhancement contributes to increased chip memory capacity and improved memory usage efficiency. Meanwhile, Intel utilizes LAB as the primary resource for implementing basic sequential and combinational circuit functions in their FPGAs [22]. The memory LAB (MLAB) represents a superset of a LAB, offering support for dual-port SRAM up to 640 bits as a distributed memory, in addition to all LAB functions. In the Hyperflex architecture by Intel, each LAB/MLAB contains 10 BLEs, with each BLE consisting of 1 adaptive 6-input LUT and 4 FFs. Notably, Intel's BLE is known as the adaptive logic module (ALM), and its structure slightly varies within LAB and MLAB.

The input/output block, arranged as groups of banks, is the interface between the FPGA and the external environment. Modern FPGAs accommodate various types of IO, broadly classified into single-ended IO (e.g., LVTTL, LVCMOS, DDR) and differential IO (e.g., LVDS, LVPECL, SerDes). Single-ended signaling, characterized by a varying voltage on a single wire, represents the simplest and most commonly employed method for transmitting electrical signals between devices. However, it is important to note that the dynamic power consumption of single-ended IO increases exponentially with the rise in clock frequency, rendering it unsuitable for high-speed circuit applications. In contrast, differential IO utilizes two wires for each signal (i.e., a differential pair).

A clock management block (CMT) is a component that manages clock functions within an FPGA. It facilitates clock synthesis, skew elimination, and clock phase and frequency adjustments. CMT enables high precision, low jitter frequency multiplication, frequency division, duty cycle adjustment, and clock phase shifting through programming. Two common implementations of CMT are the delay locked loop (DLL) and the phase locked loop (PLL). The DLL utilizes digital sampling to introduce a delay between the input and feedback clocks, ensuring consistency between their rising edges. It is also referred to as a digital phase-locked loop. On the other hand, the PLL, also known as an analog phase-locked loop, employs voltage control to manage the delay and uses a voltage-controlled oscillator (VCO) to achieve delay functionality similar to that of the DLL.

Different types of memory are available for FPGAs. On-chip memory refers to the integration of memory tile as a hard core within the FPGA. One representative of traditional on-chip memory is the block RAM (BRAM). BRAMs can be programmed as a single port, simple dual port, true dual port, read-only memory, or other modes, and the depth and width of the stored data can also be configured with high flexibility [23]. Additionally, distributed RAM configured by the LUT serves as an effective supplement to BRAM and is suitable for smaller storage requirements. Unlike BRAM, the location of distributed RAM configured by the LUT is flexible. Off-chip memory, such as DRAM, is placed on the periphery of an FPGA, providing extended storage space. External memory can also be available on the same package as that where the FPGA sits in. A typical example is high bandwidth memory

(HBM), which compactly connects stacked dynamic random access memory (DRAM) and FPGA through an interposer.

Scalar computing (SC), commonly represented by a central processing unit (CPU) and microcontroller unit (MCU), possesses unique advantages that are not achievable by traditional FPGA technology. While traditional FPGAs excel in parallel processing, scalar engines are adept at executing control operations. Combining the two can result in a higher performance-to-power ratio. It has now become a common practice to integrate SCs into FPGAs (e.g., ARM in the AMD Zynq family/RISC-V in the Microchip PolarFire family). In addition to hard core implementations, some scalar engines are designed as soft cores (e.g., Intel's NIOS/AMD's MicroBlaze).

Vector computing (VC) is typically comprised of a DSP unit and a graphics processing unit (GPU). It excels in processing a specific range of parallel computing tasks, but its performance is hindered by limited latency and efficiency attributed to the rigid nature of its memory hierarchies. Matrix computing (MC), on the other hand, is designed to provide significant performance enhancements for AI workloads, particularly in the context of matrix multiplication. In addition to the aforementioned tiles, modern FPGAs also integrate various analog or application-specific units, including but not limited to analog-to-digital/digital-to-analog conversion (ADC/DAC) units and video codec units.

Interconnect resources facilitate the connection of all CLBs and computing blocks within the FPGA through communication channels. Programmable switches traditionally govern these interconnect resources, enabling the redirection of signals along different paths. Additionally, modern FPGAs utilize network-on-chip (NoC) interconnects to enhance the internal data transfer process, functioning akin to highways. The integration of the first heterogeneous computing engine (ARM SCT) into FPGAs by Altera (acquired by Intel in 2015) in 2000 marked the commencement of the SoC FPGA era [24]. This era has witnessed successful amalgamation of various architecture types within a unified framework [25]. Notably, in 2019, Xilinx culminated this progression by consolidating nearly all prominent architectures into a single chip (Versal series), thereby propelling adaptive SoCs to new heights. Summarily, this encapsulates the evolutionary trajectory of FPGA resources.

2.3.3 Process Technology

Since the early 2000s, major FPGA manufacturers like AMD/Xilinx and Intel/Altera have been competing to advance to the most cutting-edge process node. Bulk CMOS has been the traditional technology of choice for these vendors. Silicon-on-insulator (SOI) technology utilizes SOI wafers that contain a thin insulating layer within the substrate to minimize leakage. SOI technology includes partially depleted (PD-SOI) and fully depleted (FD-SOI) types. Companies like Lattice and Quicklogic have adopted FD-SOI technology for their 28 nm and 22 nm products, respectively [26, 27].

CMOS technology introduced planar transistors in the mid-20th century. However, the downsizing of planar transistors brought numerous problems such as gate leakage currents, short channel effects, quantum tunneling leakage, variability, and mobility degradation. New technologies such as FinFET and GAAFET were then introduced to address these issues [28]. FinFET technology, such as 16 nm/14 nm nodes, features a fin-shaped body, displaying superior short-channel behavior, lower switching times, and higher current density compared to planar technology. FinFET-based FPGAs started to appear circa 2015 in relatively high-end devices.

GAAFET is a promising and futuristic transistor candidate to replace FinFET, addressing the undesirable variability and mobility loss experienced as the fin width in a FinFET approaches 5 nm. Notably, big players like Intel, Xilinx (AMD), Microchip/Microsemi, and Lattice FPGAs currently use bulk FinFET technology. Xilinx (AMD) recently announced a 7nm FPGA for the automotive industry, marking a significant advancement in the FPGA market [29].

2.3.4 Packaging Technology

The wire-bond assembly technology represents the oldest and most widely used method for IC assembly. This technology involves mounting the IC with the highest metal facing up. Small wires extend from the inputs and outputs ("I/Os") on the periphery of the IC to specific points on the package. In contrast, flip-chip has emerged as a superior alternative to wire bonds. The defining feature of the flip-chip package is the "flipped" IC, with the highest metal facing down toward the package. In a flip-chip FPGA, the interconnects are much shorter compared to wire bond technology, resulting in reduced electrical losses and heat generation.

The 2.5D package integrates multiple dies on a single interposer and interconnects those chipsets on that interposer using metal interconnects instead of the traditional planar package solutions (single die). Companies like Intel and AMD commonly adopt this type of technology for their advanced FPGA devices. True 3D package technology involves splitting FPGAs into multiple chips and stacking them. However, mass production of true 3D FPGAs is still not a reality.

2.4 Introduction to eFPGA

The distinction between FPGA and eFPGA design has become increasingly clear over the past two decades. In 2006, Menta launched the first commercially available eFPGA hard IP, enabling designers to incorporate eFPGA IP into their own ASICs seamlessly. After that, several vendors entered the market, providing commercial eFPGA IPs that can be integrated into ASIC designs [30, 31].

An eFPGA is a digitally reconfigurable structure comprising programmable logic in a programmable interconnect. It typically is deployed as a rectangular array with data inputs and outputs around the edges. eFPGAs generally feature hundreds or thousands of inputs and outputs that can be linked to buses, data paths, control paths, general purpose input/output (GPIOs), physical layer devices (PHYs), or any required device.

FPGAs have undergone advancements in process technology, packing technology, bit-stream storage, and, notably, the structure of logic tile. For instance, the Virtex 4 FPGA family employed 4-input LUTs, whereas the Virtex 5 and Virtex 6 families embraced 5-input and 6-input LUTs, respectively [20, 32]. Some manufacturers have also introduced FPGAs with larger 8-input LUTs. This progression in LUT sizes has empowered FPGAs to accommodate a wide range of intricate digital logic designs at nearly optimal configurations. In general, eFPGA macros are able to mirror the progress seen in traditional FPGAs.

eFPGAs combine the adaptability of conventional FPGAs with the integration advantages of block-based SoC designs. eFPGA integrates one or more FPGAs in IP form, serving as a general-purpose processor, coprocessor, or peripheral, typically connectable through a system bus [33]. Designers have the flexibility to tailor the eFPGA IP to meet the specific requirements of their application by specifying the number of logic units, DSP, and machine learning processing (MLP) units needed.

Development costs are increasing significantly as new process generations are introduced. These rising costs are attributed to the increasing complexity of abstract designs and the physical implementation of these designs in SoC devices, which includes software tools, engineering time, and mask costs. Conversely, the cost per unit of functionality of these devices has decreased. For instance, FPGA gates were relatively expensive two or three decades ago, resulting in their usage for prototyping and pre-production rather than mass-production applications. Attempts to incorporate FPGA gates into ASICs previously led to an increase in overall die size and complexity, making the new hybrid devices impractically expensive. However, today, directly integrating programmable logic into custom chips presents various advantages for diverse applications.

This approach offers enhanced flexibility, cost reduction, and a smaller eFPGA IP footprint by excluding unnecessary FPGA features. Moreover, if a custom or specialized FPGA architecture is required, it can also be implemented as an eFPGA for improved performance. As described previously, multiple IP providers offer eFPGA blocks with different granularities and architectures [34]. These IPs are licensable and are used similarly to other IPs in semiconductor design.

The illustration in Fig. 2.5 demonstrates the difference between FPGA and eFPGA. SoC design expenses are significantly higher than FPGA, which increases the risk of not meeting specific market demands with the right product. The use of eFPGA-based SoC design makes a significant contribution. Additionally, an important aspect of eFPGA, as shown in Fig. 2.5, is that part of the design implemented on the eFPGA is not present until the bitstream is loaded. This reconfigurability feature conceals the design. An example to better understand

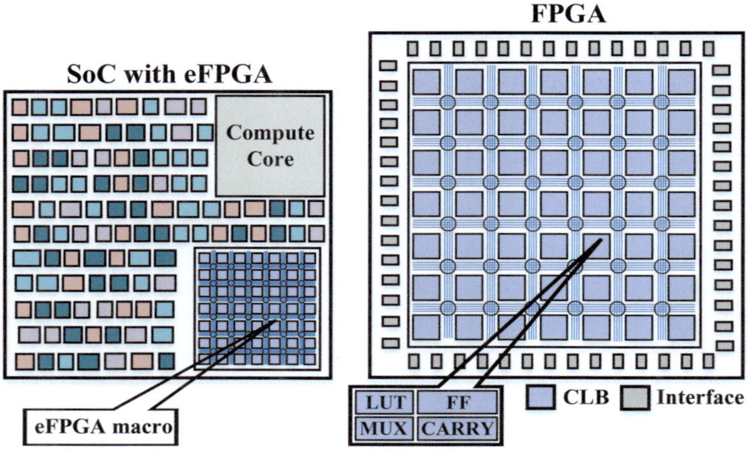

Fig. 2.5 Difference between FPGA and eFPGA-based device

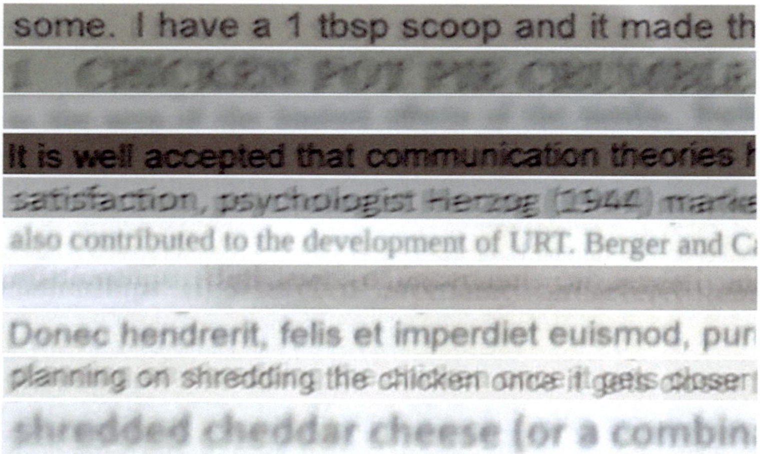

Fig. 2.6 An example for the layman of hidden text could be compared to obfuscation

this concept is illustrated in Fig. 2.6. This is akin to a document with some lines, some fully visible, some partially visible, and some completely hidden. In the context of using eFPGA for obfuscation, this concept is similar to Fig. 2.6. Part of the eFPGA design is completely hidden, while the remaining part is visible. From an attacker's perspective, significant effort may result in obtaining partial information, akin to the partially visible lines in the example.

2.5 Takeaway Notes

In this chapter, we introduced different hardware devices and explained how the architecture of the FPGA evolved over time. We discussed how reconfigurability is achieved through the FPGA and how an ASIC can leverage this feature with the help of eFPGA, which is also introduced in this chapter. We have also introduced the concept of obfuscation and explained how eFPGA offers obfuscation in simple terms. In Chap. 3, we will explain the IC design flow with ReBO and provide details about conceptual obfuscation using ReBO.

References

1. K. Bob, "How hybrid structured asics provide low cost solutions for mid-range applications." https://www.eetimes.com/how-hybrid-structured-asics-provide-low-cost-solutions-for-mid-range-applications/, 2006. Accessed: August 15, 2024.
2. S.-C. Chen and Y.-W. Chang, "Fpga placement and routing," in *2017 IEEE/ACM International Conference on Computer-Aided Design (ICCAD)*, pp. 914–921, 2017.
3. Invent Logic, "FPGA architecture," last accessed on Mar 30, 2023. Available at: https://allaboutfpga.com/fpga-architecture/.
4. EDN, "How to design an fpga architecture tailored for efficiency and performance." https://www.eetimes.com/how-to-design-an-fpga-architecture-tailored-for-efficiency-and-performance/, 2007. Accessed: June 20, 2024.
5. H. Krupnova, C. Rabedaoro, and G. Saucier, "Synthesis and floorplanning for large hierarchical fpgas," in *Proceedings of the 1997 ACM Fifth International Symposium on Field-Programmable Gate Arrays*, FPGA '97, p. 105–111, 1997.
6. R. Z. Chochaev and P. I. Frolova, "Initial placement algorithms for island-style fpgas," in *2022 Conference of Russian Young Researchers in Electrical and Electronic Engineering (ElConRus)*, pp. 586–589, 2022.
7. B. Santo, "25 microchips that shook the world." https://spectrum.ieee.org/25-microchips-that-shook-the-world, 2009. Accessed: June 20, 2024.
8. I. Spectrum, "Chip hall of fame: Xilinx xc2064 fpga hardware that can transform itself on command has proven incredibly useful." https://spectrum.ieee.org/chip-hall-of-fame-xilinx-xc2064-fpga, 2017. Accessed: June 19, 2024.
9. M. Abusultan and S. P. Khatri, "Exploring static and dynamic flash-based fpga design topologies," in *2016 IEEE 34th International Conference on Computer Design (ICCD)*, pp. 416–419, 2016.
10. N. Rezzak, J.-J. Wang, D. Dsilva, and N. Jat, "Tid and see characterization of microsemi's 4th generation radiation tolerant rtg4 flash-based fpga," in *2015 IEEE Radiation Effects Data Workshop (REDW)*, pp. 1–6, 2015.
11. P. R. C. Villa, R. C. Goerl, F. Vargas, L. B. Poehls, N. H. Medina, N. Added, V. A. P. de Aguiar, E. L. A. Macchione, F. Aguirre, M. A. G. da Silveira, and E. A. Bezerra, "Analysis of single-event upsets in a microsemi proasic3e fpga," in *2017 18th IEEE Latin American Test Symposium (LATS)*, pp. 1–4, 2017.
12. J. Greene, S. Kaptanoglu, W. Feng, V. Hecht, J. Landry, F. Li, A. Krouglyanskiy, M. Morosan, and V. Pevzner, "A 65nm flash-based fpga fabric optimized for low cost and power," in *Proceedings of the 19th ACM/SIGDA International Symposium on Field Programmable Gate Arrays*, p. 87–96, Association for Computing Machinery, 2011.

13. O. Turkyilmaz, S. Onkaraiah, M. Reyboz, F. Clermidy, C. A. Hraziia, J. Portal, and M. Bocquet, "Rram-based fpga for "normally off, instantly on" applications," in *2012 IEEE/ACM International Symposium on Nanoscale Architectures* (NANOARCH), pp. 101–108, 2012.
14. S. Ikegawa, F. B. Mancoff, J. Janesky, and S. Aggarwal, "Magnetoresistive random access memory: Present and future," *IEEE Transactions on Electron Devices*, vol. 67, no. 4, pp. 1407–1419, 2020.
15. X. Tang, E. Giacomin, G. De Micheli, and P.-E. Gaillardon, "Circuit designs of high-performance and low-power rram-based multiplexers based on 4t(ransistor)1r(ram) programming structure," *IEEE Transactions on Circuits and Systems I: Regular Papers*, vol. 64, no. 5, pp. 1173–1186, 2017.
16. B. Govoreanu, G. Kar, Y.-Y. Chen, V. Paraschiv, S. Kubicek, A. Fantini, I. Radu, L. Goux, S. Clima, R. Degraeve, N. Jossart, O. Richard, T. Vandeweyer, K. Seo, P. Hendrickx, G. Pourtois, H. Bender, L. Altimime, D. Wouters, J. Kittl, and M. Jurczak, "10×10nm2 hf/hfox crossbar resistive ram with excellent performance, reliability and low-energy operation," in *2011 International Electron Devices Meeting*, pp. 31.6.1–31.6.4, 2011.
17. X. Tang, G. Kim, P.-E. Gaillardon, and G. De Micheli, "A study on the programming structures for rram-based fpga architectures," *IEEE Transactions on Circuits and Systems I: Regular Papers*, vol. 63, no. 4, pp. 503–516, 2016.
18. J. Cong and B. Xiao, "Fpga-rpi: A novel fpga architecture with rram-based programmable interconnects," *IEEE Transactions on Very Large Scale Integration (VLSI) Systems*, vol. 22, no. 4, pp. 864–877, 2014.
19. R. Rajaei, "Radiation-hardened design of nonvolatile mram-based fpga," *IEEE Transactions on Magnetics*, vol. 52, no. 10, pp. 1–10, 2016.
20. Xilinx Inc., "Virtex-6 family overview," last accessed on Jan 09, 2023. Available at: https://docs.xilinx.com/v/u/en-US/ds150.
21. X. (AMD), "Versal acap configurable logic block architecture manual (am005)." https://docs.amd.com/r/en-US/am005-versal-clb/CLB-Architecture, 2023. Accessed: June 20, 2024.
22. Intel, "Intel stratix 10 logic array blocks and adaptive logic modules user guide." https://docs.amd.com/r/en-US/am005-versal-clb/CLB-Architecture, 2023. Accessed: June 20, 2024.
23. Efinix, "Block ram wrapper core." https://www.efinixinc.com/support/ip/bram-wrapper.php, 2024. Accessed: August 15, 2024.
24. Xilinx, "Xilinx ships over 10 million platform flash memory devices." https://www.design-reuse.com/news/13990/xilinx-10-million-platform-flash-memory-devices.html, 2006. Accessed: February 10, 2023.
25. A. Boutros and V. Betz, "Fpga architecture: Principles and progression," *IEEE Circuits and Systems Magazine*, vol. 21, no. 2, pp. 4–29, 2021.
26. Samsung, "Fd-soi, the disruptive innovation samsung foundry is leading to overcome the limits." https://semiconductor.samsung.com/news-events/tech-blog/fd-soi-the-disruptive-innovation-samsung-foundry-is-leading-to-overcome-the-limits/, 2023. Accessed: June 21, 2024.
27. Lattice, "Lattice nexus platform: Enabling low power, high reliability, and high performance design." https://www.latticesemi.com/LatticeNexus, 2019. Accessed: June 21, 2024.
28. T. Matt, "Introduction to gaafet: The next big phase of computer chip manufacturing." https://medium.com/predict/introduction-to-gaafet-the-next-big-phase-of-computer-chip-manufacturing-84e63abe11dd, 2023. Accessed: June 21, 2024.
29. F. Nick, "Amd shows first 7nm automotive fpga and embedded processor." https://www.eenewseurope.com/en/amd-shows-first-7nm-automotive-fpga-and-embedded-processor/, 2024. Accessed: June 21, 2024.

30. Menta, "Embedded FPGA IP," last accessed on Apr 2, 2022. Available at: https://www.menta-efpga.com/.
31. Achronix Corp., "Speedcore embedded FPGA IP," last accessed on Apr 22, 2023. Available at: https://www.achronix.com/product/speedcore.
32. Xilinx Inc., "Virtex-5 family overview," last accessed on Jan 09, 2023. Available at: https://docs.xilinx.com/v/u/en-US/ds100.
33. Design & Reuse, "Menta unveils new efpga technology." https://www.design-reuse.com/news/12774/menta-efpga-technology.html, 2006. Accessed: March 12, 2023.
34. Achronix Inc., "Speedster7t fpgas," last accessed on Apr 18, 2022. Available at: https://www.achronix.com/high-speed-interfaces.

ReBO-Driven IC Design: Leveraging Reconfigurable Logic for Obfuscation

3

3.1 Secure IC Design Flow

The secure design flow with ReBO is illustrated in Fig. 3.1. In the CAD flow, there is an additional step where a custom tool reads the netlist and applies user-defined security criteria to obfuscate the design. The user-defined security criteria contain various constraints, such as the target obfuscation level and the targeted performance. Optionally, the critical parts of the design might be annotated to indicate where the obfuscation should be prioritized. The designer implements the hardware description language (HDL) according to the design specifications. In the next stage, a custom obfuscation tool obfuscates the design, and then the obfuscated netlist is synthesized with a commercial logic synthesis tool. After that, the designer observes the PPA overheads and adjusts the obfuscation criteria accordingly. The designer repeats this step until the targeted security criteria are met. These criteria could have multiple options depending on the design and requirements. Some of the high-level possible criteria could be described as follows:

- The designer wants the maximum security (obfuscation) possible without compromising performance.
- The designer wants the minimum security necessary to protect against a given adversarial capability (e.g., 1024-bit long keys).
- The designer gives a budget for performance degradation (e.g., 5%) and seeks the maximum security allowed under that budget.
- The designer wants half of the logic to be obfuscated.

After the obfuscation, the design represents a hybrid of reconfigurable and non-reconfigurable logic (i.e., static logic). After logic synthesis, the design is subjected to physical synthesis to generate a layout. The secure IC design methodology could be differ-

Z. U. Abideen and S. Pagliarini, *Reconfigurable Obfuscation Techniques for the IC Supply Chain*, Synthesis Lectures on Digital Circuits & Systems,
https://doi.org/10.1007/978-3-031-77509-3_3

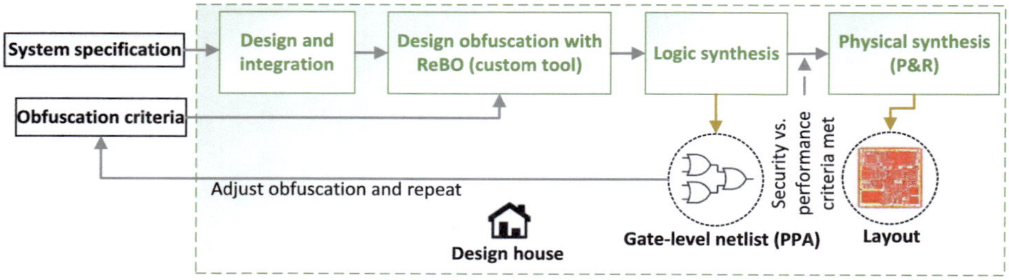

Fig. 3.1 Secure IC design flow with ReBO

ent based on the class of the ReBO. For example, in the case of eFPGA hard IP, extra effort is required at the floorplan stage. All these classifications will be discussed in detail in the upcoming chapters.

3.2 An Example of Obfuscation with ReBO

This section provides a theoretical explanation of ReBO. It introduces common terminologies for ReBO.

The initial netlist is secured by a bitstream known exclusively by the IP owner. This secured netlist undergoes untrusted design processes, including an untrusted foundry, an outsourced test facility, and a deployment phase. Due to the lack of access to the bitstream, unauthorized entities cannot exploit the design's accurate functionality. To activate a fabricated IC, the IP owner loads the bitstream in a trusted environment.

Figure 3.2 demonstrates an obfuscated netlist incorporating 2-input LUTs. Wires from the original design become inputs to the LUTs. The other LUT inputs, denoted as configuration bits, are stored in an NVM [1]. An obfuscated IC (or locked netlist) is incapable of generating

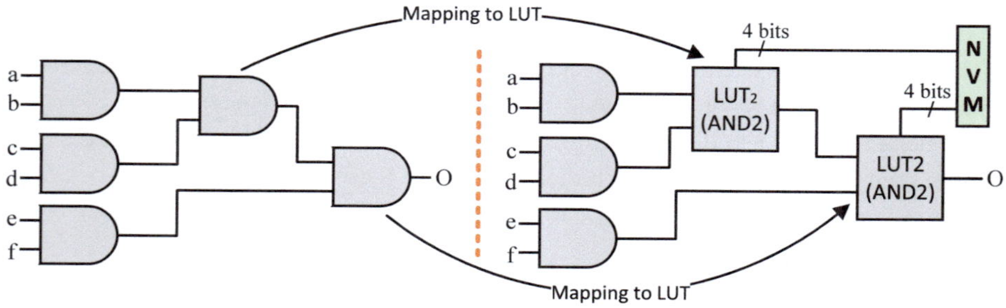

Fig. 3.2 Design obfuscation with LUTs: an example of circuit with six inputs (left), and an obfuscated circuit locked with 2 LUTs and correct configuration bits with "00010001" (right)

the correct output unless it is activated using the appropriate bitstream. Considering the obfuscated circuit in Fig. 3.2, the gates are mapped to 2-input LUTs that can implement four different functions. For the AND gate, loading "0001" as the configuration bits for the LUT to the NVM allows all the LUTs to implement the functions and produce the correct output for all input patterns. For instance, the locked circuit outputs O = 1 only when the input pattern is "11 11 11". However, applying an incorrect key value, such as "01110110", results in certain gates behaving as other Boolean functions, leading to an injection of errors in the circuit when the same input pattern is applied. Evidently, the error injected into the circuit is inconsistent across the input patterns. While the given example is rather small, it is easy to imagine that, for a representative circuit, the adversary has no means of evaluating all possible input patterns against all possible key guesses.

3.3 ReBO Versus Other Countermeasure Techniques

The techniques mentioned in Chap. 1 aim to primarily provide security against threats related to IC fabrication. The various techniques differ in terms of the threat model and security objectives. The threat model defines the trusted or untrusted entities and the assets to which each entity has access. A threat model is a blue print for defining the capabilities of an adversary. Table 3.1 compares existing countermeasures and their efficiency against attacks mounted on different stages of the IC design process or the IC supply chain.

Camouflaging can only protect the design during the deployment stage. Split manufacturing provides protection during fabrication and deployment stages, but it relies partially on a trusted foundry. Watermarking and fingerprinting can only detect IP piracy or theft after it has occurred and do not prevent it. Passive metering offers protection during the testing, packaging, and deployment stages. LL can protect against security threats at various stages in the design process, with the exception of the trusted design house. LL is a well-researched

Table 3.1 The security of countermeasures against attacks mounted at various stages of the globalized IC supply chain

Countermeasure technique	System on chip (SoC) integrator	Fabrication	Test and Packaging	Deployment
Camouflaging [6–8]	✗	✗	✗	✔
Split manufacturing [9, 10]	✗	✔	✗	✗
Metering (passive) [11–13]	✗	✗	✔	✔
Logic locking [14–18]	✔	✔	✔	✔
ReBO [1, 19–22]	✔	✔	✔	✔

Prevents attack? ✔ (Yes), ✗ (No)

topic and has gained attention due to its potential to offer security. There are many variants of LL in existence today [2–5]. LL does not require foundry support like camouflaging and does not require trusted BEOL foundry like split manufacturing.

Significant concerns exist regarding attacks on camouflaging, split manufacturing, and LL. For instance, removal and boolean satisfiability (SAT) attacks on camouflaging could fully or partially deobfuscate the design [6, 23]. Attacks on split manufacturing are also prevalent, as shown in [24, 25]. While LL provides high security at all stages of the IC supply chain, the SAT attack on LL broke its security. The initial SAT attack compromised the security measures implemented by LL [26]. After the SAT attack, the security of LL became more prevalent, and then various agencies started to fund this research lien, including the DARPA [27]. Numerous SAT attacks have been proposed on LL [14–18]. LL has recently seen a *rise in the interplay between the countermeasure and attack techniques*. Still, it has been observed that LL is vulnerable to numerous attacks [26, 28–33].

On the contrary, ReBO techniques have shown significant potential in providing quality obfuscation that can withstand various security threats. ReBO has the same potential as LL, and an active research community is working on it. These techniques employ various types of reconfigurable elements, such as SRAM-based LUTs [34–38], FF-based LUTs [1, 39] and NVM-based LUTs [40–46]. Additionally, other ReBO approaches have been introduced in recent years to enhance the protection of digital designs [19–22, 47–54]. Few other approaches involve configuring transistors and switches instead of LUTs, as described in [21, 49]. Most of the ReBO techniques use LUTs as valuable assets in securing the integrity of a design.

3.3.1 ReBO Versus LL

LL and ReBO techniques share the aim of protecting IP against supply chain attacks. ReBO has gained increased attention for its high resiliency against state-of-the-art attacks, but concerns have arisen regarding the trade-offs between security and PPA. Figure 3.3 represents the conceptual difference between LL and a typical eFPGA redaction approach. The eFPGA solution on the right side of the figure is purposefully bigger to highlight the PPA trade-off associated with ReBO techniques.

As previously mentioned, LL involves adding gates with key inputs to the original design. It also requires the correct configuration of the secret key in a tamper-proof memory. In contrast, ReBO relies on loading the correct bitstream and utilizes reconfigurable logic elements instead of keyed gates. These approaches exhibit some similarities, and as a result, key finding attacks developed for the LL attacks can also be applied to ReBO techniques. Considering the brute force, the search space of each reconfigurable in ReBO is 2^m, where m is the number of configuration bits for a single reconfigurable element. ReBO creates a large bitstream or key bits with reconfigurable elements. It is essential to note that a large bitstream creates a vast search space with 2^n key combinations where n represents the number of key inputs. This leads to an exponential increase in the search space for the correct key, making the brute force attack computationally infeasible.

In addition to the brute force attack, LL is also vulnerable to various other attacks, such as structural attacks, algorithmic attacks, and side-channel attacks [6, 55, 56]. As a result, the design's overall security may still be compromised even when obfuscation is applied. The adversary gains access to the entire design, comprising the original IP and the key gates. If the adversary has determined the key values, he/she can recreate the design by setting the keys to a constant and removing the tamper-proof memory. This becomes much more difficult to achieve within a ReBO solution since the eFPGA macro is, in fact, part of the original design. It cannot be easily replaced or swapped out.

3.3.2 Need of Custom CAD Tool

In the domain of LL, designers use the conventional CAD design flow to leverage various tools for logic synthesis, timing analysis, and design optimization. The process of locking a design tends to be carried out with simple scripts that take a netlist as input and also produce a netlist as output. Thus, the conventional flow is not disrupted.

On the other hand, ReBO lacks a compatible tool with the conventional CAD flow that can support these functionalities [47]. Figure 3.3 demonstrates an instance of ReBO that exploits eFPGA to lock the design. The ASIC part of an eFPGA-based obfuscation or eFPGA redaction is considered to be static. Hence, it is referred to as the non-reconfigurable part. Every ReBO technique *employs a different methodology* for selecting what logic becomes reconfigurable and what logic remains static. ReBO solutions have attempted partitioning the design on the high-level synthesis (HLS) level or partitioning on the HDL level. Therefore,

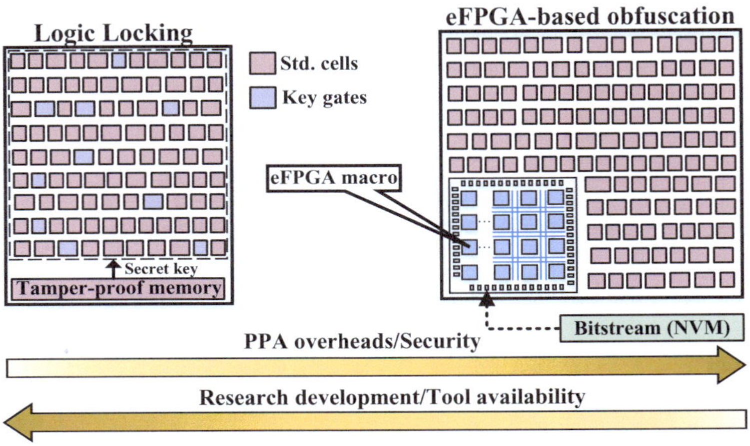

Fig. 3.3 Comparison diagram: LL versus eFPGA-based obfuscation

developing customized tools is necessary to implement such techniques, which requires significant effort. LL benefits from more sophisticated techniques and readily available tools, as indicated by the arrows pointing towards the left in Fig. 3.3. Nevertheless, ReBO results in higher security that is, in turn, accompanied by PPA overheads.

In conclusion, it can be stated that LL techniques are more established and supported by readily available tools, including open sourced alternatives [57, 58]. At the same time, reconfigurable-based obfuscation shows promise but requires further development to overcome existing hurdles.

3.3.3 Need of Robust Optimization Techniques

For efficient utilization of reconfigurable logic, it becomes necessary to consider the PPA overheads along with security/obfuscation goals. ReBO techniques face several challenges during the design, fabrication, testing, and deployment stages. For example, the SRAM-based LUT implementation requires careful consideration of the placement of SRAM. On the other hand, MRAM designed with STT-magnetic tunnel junction (MTJ) and SOT-MTJ are hybrid and emerging technologies that require specific considerations and capabilities during the fabrication process—there are very few foundries that can offer such emerging devices.

However, incorporating reconfigurable elements into the design introduces a certain level of PPA overhead, which may affect the overall performance of the design. During the design phase, PPA overheads are important and remain so during obfuscation. For this reason, most of the ReBO works suggest selecting a small yet vital portion of the design to place on the reconfigurable logic. As advancements in ASIC designs progress, balancing robust security measures with high performance and low area utilization proves to be a constant challenge for designers. For ReBO, the most sensitive part of the design is redacted to the eFPGA, while the remaining portion or non-reconfigurable part uses standard cells [19]. In both cases, either LUT-based obfuscation or eFPGA redaction requires *storing the bitstream* of the design. The aforementioned techniques utilize SRAM, FF, and NVM, such as MRAM designed with STT-MTJ and SOT-MTJ technology to store the bitstream.

3.4 Security of ReBO Against Supply Chain Threats

This section will cover how ReBO stays protected from potential supply chain threats. We will discuss the traditional threat model and the main security strengths of ReBO.

3.4.1 Threat Model

In terms of the threat model, when considering the untrusted foundry, we do not differentiate between whether the adversary is institutional or a single rogue employee. We assume that the security of ReBO relies on its non-reconfigurable and reconfigurable logic, which is secured by a bitstream. Based on this, we present the following assumptions:

- The primary objective of the adversary is to analyze the design with the intention of replicating its IP illegally, fabricating a larger than authorized quantity of ICs, or covertly inserting hardware trojans. In pursuit of this objective, the adversary is compelled to obtain and reconstruct the bitstream.
- The objective of the adversary may include the identification of the high-level circuit intent, even when obfuscation is present. In pursuit of this objective, the adversary is not required to reconstruct the bitstream.
- The adversary possesses the GDSII file of the design that was submitted for manufacturing. The adversary is adept in IC design and possesses the necessary expertise and resources to understand this particular layout representation.
- The adversary has the capability to identify the standard cells, which means that the gate-level netlist of the obfuscated circuit is readily retrieved [59].
- The adversary is capable of identifying the reconfiguration pins [3, 26], allowing for the easy enumeration of all reconfigurable elements and their configuration sequence.
- The adversary can isolate the reconfigurable and non-reconfigurable logic by first recovering the netlist and the configuration order of the reconfigurable elements.
- The adversary may gain insights into the structural characteristics of the reconfigurable logic, allowing them to potentially reconstruct portions of the bitstream.

3.4.2 Security Characteristics of ReBO

Concerning **reverse engineering and IP piracy**, the adversary can apply the same physical reverse engineering approaches that would be used on a conventional ASIC. Even if the process is long and error prone, there is no characteristic of a ReBO design that can prevent the act of reverse engineering. However, the comprehension of what the reverse engineered logic does is what is taken from the adversary. The adversary will be able identify the reconfigurable and non-reconfigurable parts, but cannot directly infer what bitstream to be used for configuring the obfuscated portion.

In other words, without the knowledge of the correct bitstream, the extracted netlist will not fully correspond to the IC and will be of little use to the adversary. The reconfigurable elements, such as LUTs, represent many potential logic functions, thus introducing ambiguity in design reconstruction.

In the case of **overbuilding**, the adversary is an untrusted foundry interested in producing additional chips to sell them in the grey market at lower prices. However, the fabricated ICs cannot be activated without the bitstream and remain non-functional.

ReBO also helps to prevent the insertion of **hardware trojans** in a circuit by obfuscating parts of the circuit. For a trojan to be successful, the adversary must understand the functionality of the design and then place the trojan to achieve that goal. For instance, if the adversary is attempting to leak private data, he/she must understand where the data is available within the circuit. In the case of ReBO, only the non-reconfigurable part is exposed to the adversary, making it harder for the attacker to identify candidate locations for inserting trojans. From the structural point of view, the reconfigurable part is a regular grid in the case of eFPGA redaction. Even in the case of other ReBo techniques, every reconfigurable part is somewhat similar if, for instance, same-sized LUTs are employed. Thus, the adversary would be left with the ability to insert a random trojan, which is very unlikely to be of any use.

In the context of **counterfeiting**, ReBO is designed with the primary objective of securing against piracy. To achieve this, ReBO utilizes a bitstream protection mechanism integrated with a PUF that generates a secret key for the cryptographic core embedded within it. The PUF serves as a potential means of authentication by leveraging a large dataset of PUF responses to verify the authenticity of ReBO. This approach aligns with prior research that has successfully employed PUF for IC authentication [60]. Using PUF for security can help make ReBO more secure against counterfeiting and piracy.

3.5 Security at End-User Stage

Once an IC containing a ReBO approach is deployed, the security of the bitstream becomes crucial as it contains the secret to unlock the most sensitive part of the design. It has been shown that many attacks are capable of reverse engineering the bitstream of FPGAs or reconfigurable hardware [61–65]. These days, modern FPGA devices are equipped with encryption and authentication techniques. For instance, the Xilinx Vivado Design Suite supports Advanced Encryption Standard (AES) and Rivest "Shamir" Adleman (RSA)-based authentication modes [62]. AES is a widely recognized cryptographic standard of the National Institute of Standards and Technology (NIST) and the U.S. Department of Commerce [66]. This ensures the bitstream remains unmodified and can only be deciphered using a dedicated on-chip decryption block.

Physical unclonable functions (PUFs) are promising primitives for cryptography and hardware security in general. PUFs can be used to generate a unique secret key (or a seed for a key) that is then used together with AES or RSA [67, 68]. Additionally, their non-reproducible and unclonable properties result in the production of a unique signature per chip. PUFs can too be classified into different groups, including ring-oscillator based PUF (RO-PUF) [69], arbiter PUF [70], dynamic random access memory (DRAM) PUF [71], and SRAM-based PUF [72–80]. SRAM-based PUFs, in particular, offer a combination of simplicity, low cost, high reliability, scalability, and cryptographic strength, making them a popular choice for commercial PUF solutions [67]. Additionally, they rely on standard SRAM IP, which is readily available to designers and eliminates the need for IP customization.

Side-channel attacks also pose a significant threat originating from end-users, where the ReBO is deployed in the IC-based system, allowing the user to exploit the current and power profiles of the device to gather information about the secret. This especially concerns a cryptography core running on a ReBO solution. Generally, side-channel attacks on the ReBO can be categorized into two major types. The first type of attack occurs when the ReBO is running a cryptographic core, and the adversary aims to exploit the structure of the crypto core, along with certain hypotheses, to recover the key of the crypto core [63, 81–84]. The second class of attack is executed with the goal of recovering the bitstream of the ReBO itself, irrespective of the design that it implements. In the context of FPGA architectures, various side-channel attacks of both classes have been proposed. In principle, ReBO approaches are susceptible to the same attacks but perhaps not to the same degree—ReBO may offer help in mitigating side-channel attacks [85].

3.6 Takeaway Notes

In this chapter, we presented a secure IC design flow with ReBO. The chapter discusses a detailed comparison of ReBO with other countermeasure techniques. Additionally, it discussed different aspects associated with ReBO, such as the development of custom tools, the need for optimization techniques, and security at the end-user stage. So far, we have discussed the ReBO and its performance and security compared to LL.

References

1. Z. U. Abideen, T. D. Perez, M. Martins, and S. Pagliarini, "A security-aware and lut-based cad flow for the physical synthesis of hasics," *IEEE Transactions on Computer-Aided Design of Integrated Circuits and Systems*, vol. 42, no. 10, pp. 3157–3170, 2023.
2. M. Yasin, B. Mazumdar, J. J. V. Rajendran, and O. Sinanoglu, "Sarlock: Sat attack resistant logic locking," in *2016 IEEE International Symposium on Hardware Oriented Security and Trust (HOST)*, pp. 236–241, 2016.
3. M. Yasin, A. Sengupta, M. T. Nabeel, M. Ashraf, J. J. Rajendran, and O. Sinanoglu, "Provably-secure logic locking: From theory to practice," in *Proceedings of the 2017 ACM SIGSAC Conference on Computer and Communications Security*, p. 1601–1618, 2017.
4. Y. Xie and A. Srivastava, "Anti-sat: Mitigating sat attack on logic locking," *IEEE Transactions on Computer-Aided Design of Integrated Circuits and Systems*, vol. 38, no. 2, pp. 199–207, 2019.
5. K. Shamsi, M. Li, T. Meade, Z. Zhao, D. Z. Pan, and Y. Jin, "Cyclic obfuscation for creating sat-unresolvable circuits," in *Proceedings of the Great Lakes Symposium on VLSI 2017*, GLSVLSI '17, (New York, NY, USA), p. 173–178, Association for Computing Machinery, 2017.

6. M. Yasin, B. Mazumdar, O. Sinanoglu, and J. Rajendran, "Removal attacks on logic locking and camouflaging techniques," *IEEE Transactions on Emerging Topics in Computing*, vol. 8, no. 2, pp. 517–532, 2020.
7. R. P. Cocchi, J. P. Baukus, L. W. Chow, and B. J. Wang, "Circuit camouflage integration for hardware IP protection," in *2014 51st ACM/EDAC/IEEE Design Automation Conference (DAC)*, pp. 1–5, 2014.
8. M. Li, K. Shamsi, T. Meade, Z. Zhao, B. Yu, Y. Jin, and D. Z. Pan, "Provably secure camouflaging strategy for IC protection," *IEEE Transactions on Computer-Aided Design of Integrated Circuits and Systems*, vol. 38, no. 8, pp. 1399–1412, 2019.
9. T. D. Perez and S. Pagliarini, "A survey on split manufacturing: Attacks, defenses, and challenges," *IEEE Access*, vol. 8, pp. 184013–184035, 2020.
10. J. Rajendran, O. Sinanoglu, and R. Karri, "Is split manufacturing secure?," in *2013 Design, Automation Test in Europe Conference Exhibition (DATE)*, pp. 1259–1264, 2013.
11. Y. M. Alkabani and F. Koushanfar, "Active hardware metering for intellectual property protection and security," in *Proceedings of 16th USENIX Security Symposium on USENIX Security Symposium*, SS'07, (USA), USENIX Association, 2007.
12. F. Koushanfar, "Integrated circuits metering for piracy protection and digital rights management: An overview," in *Proceedings of the 21st Edition of the Great Lakes Symposium on Great Lakes Symposium on VLSI*, GLSVLSI '11, (New York, NY, USA), p. 449–454, Association for Computing Machinery, 2011.
13. F. Koushanfar, "Hardware metering: A survey," in *Introduction to Hardware Security and Trust* (M. Tehranipoor and C. Wang, eds.), pp. 103–122, New York, NY: Springer New York, 2012.
14. K. Zamiri Azar, H. Mardani Kamali, H. Homayoun, and A. Sasan, "Threats on logic locking: A decade later," in *GLSVLSI '19: Proceedings of the 2019 on Great Lakes Symposium on VLSI*, p. 471–476, 2019.
15. Y. Xie and A. Srivastava, "Delay locking: Security enhancement of logic locking against IC counterfeiting and overproduction," in *2017 54th ACM/EDAC/IEEE Design Automation Conference (DAC)*, pp. 1–6, 2017.
16. M. Yasin, J. Rajendran, and O. Sinanoglu, "Trustworthy hardware design: Combinational logic locking techniques," Springer, Cham, 2019.
17. M. Yasin and O. Sinanoglu, "Evolution of logic locking," in *2017 IFIP/IEEE International Conference on Very Large Scale Integration (VLSI-SoC)*, pp. 1–6, 2017.
18. J. Sweeney, V. Mohammed Zackriya, S. Pagliarini, and L. Pileggi, "Latch-based logic locking," in *2020 IEEE International Symposium on Hardware Oriented Security and Trust (HOST)*, pp. 132–141, 2020.
19. B. Hu, T. Jingxiang, S. Mustafa, R. R. Gaurav, S. William, M. Yiorgos, C. S. Benjamin, and S. Carl, "Functional obfuscation of hardware accelerators through selective partial design extraction onto an embedded FPGA," in *Proceedings of the 2019 Great Lakes Symposium on VLSI*, p. 171–176, 2019.
20. J. Chen, M. Zaman, Y. Makris, R. D. S. Blanton, S. Mitra, and B. C. Schafer, "DECOY: Deflection-Driven HLS-Based Computation Partitioning for Obfuscating Intellectual PropertY," in *Proceedings of the 57th ACM/EDAC/IEEE Design Automation Conference*, DAC '20, IEEE Press, 2020.
21. M. M. Shihab, J. Tian, G. R. Reddy, B. Hu, W. Swartz, B. Carrion Schaefer, C. Sechen, and Y. Makris, "Design obfuscation through selective post-fabrication transistor-level programming," in *2019 Design, Automation & Test in Europe Conference & Exhibition (DATE)*, pp. 528–533, 2019.
22. J. Bhandari, A. K. Thalakkattu Moosa, B. Tan, C. Pilato, G. Gore, X. Tang, S. Temple, P.-E. Gaillardon, and R. Karri, "Exploring eFPGA-based redaction for IP protection," in *2021 IEEE/ACM International Conference On Computer Aided Design (ICCAD)*, pp. 1–9, 2021.

23. M. El Massad, S. Garg, and M. V. Tripunitara, "The SAT attack on IC camouflaging: Impact and potential countermeasures," *IEEE Transactions on Computer-Aided Design of Integrated Circuits and Systems*, vol. 39, no. 8, pp. 1577–1590, 2020.
24. W. Zeng, B. Zhang, and A. Davoodi, "Analysis of security of split manufacturing using machine learning," *IEEE Transactions on Very Large Scale Integration (VLSI) Systems*, vol. 27, no. 12, pp. 2767–2780, 2019.
25. S. Chen and R. Vemuri, "On the effectiveness of the satisfiability attack on split manufactured circuits," in *2018 IFIP/IEEE International Conference on Very Large Scale Integration (VLSI-SoC)*, pp. 83–88, 2018.
26. P. Subramanyan, S. Ray, and S. Malik, "Evaluating the security of logic encryption algorithms," in *2015 IEEE International Symposium on Hardware Oriented Security and Trust (HOST)*, pp. 137–143, 2015.
27. DARPA, "Expanding domestic manufacturing of secure, custom chips for defense needs." https://www.darpa.mil/news-events/2021-03-18, 2024. Accessed: August 16, 2024.
28. J. Rajendran, Y. Pino, O. Sinanoglu, and R. Karri, "Security analysis of logic obfuscation," in *Design Automation Conference*, pp. 83–89, 2012.
29. H. Zhou, R. Jiang, and S. Kong, "CycSAT: SAT-based attack on cyclic logic encryptions," in *2017 IEEE/ACM International Conference on Computer-Aided Design (ICCAD)*, pp. 49–56, 2017.
30. A. Mondal, M. Zuzak, and A. Srivastava, "StatSAT: A boolean satisfiability based attack on logic-locked probabilistic circuits," in *2020 57th ACM/IEEE Design Automation Conference (DAC)*, pp. 1–6, 2020.
31. N. Limaye, S. Patnaik, and O. Sinanoglu, "Valkyrie: Vulnerability assessment tool and attack for provably-secure logic locking techniques," *IEEE Transactions on Information Forensics and Security*, vol. 17, pp. 744–759, 2022.
32. S. Patnaik, N. Limaye, and O. Sinanoglu, "Hide and seek: Seeking the (un)-hidden key in provably-secure logic locking techniques," *IEEE Transactions on Information Forensics and Security*, vol. 17, pp. 3290–3305, 2022.
33. Z. U. Abideen, S. Gokulanathan, M. J. Aljafar, and S. Pagliarini, "An overview of FPGA-inspired obfuscation techniques," *Association for Computing Machinery*, vol. 56, no. 12, December 2024.
34. G. Kolhe, T. Sheaves, K. I. Gubbi, T. Kadale, S. Rafatirad, S. M. PD, A. Sasan, H. Mahmoodi, and H. Homayoun, "Silicon validation of LUT-based logic-locked IP cores," in *Proceedings of the 59th ACM/IEEE Design Automation Conference*, pp. 1189–1194, 2022.
35. S. D. Chowdhury, G. Zhang, Y. Hu, and P. Nuzzo, "Enhancing SAT-attack resiliency and cost-effectiveness of reconfigurable-logic-based circuit obfuscation," in *2021 IEEE International Symposium on Circuits and Systems (ISCAS)*, pp. 1–5, IEEE, 2021.
36. A. Baumgarten, A. Tyagi, and J. Zambreno, "Preventing IC piracy using reconfigurable logic barriers," *IEEE Design Test of Computers*, vol. 27, no. 1, pp. 66–75, 2010.
37. B. Liu and B. Wang, "Embedded reconfigurable logic for ASIC design obfuscation against supply chain attacks," in *2014 Design, Automation Test in Europe Conference Exhibition (DATE)*, pp. 1–6, 2014.
38. H. Mardani Kamali, K. Zamiri Azar, K. Gaj, H. Homayoun, and A. Sasan, "LUT-lock: A novel LUT-based logic obfuscation for FPGA-bitstream and ASIC-hardware protection," in *2018 IEEE Computer Society Annual Symposium on VLSI (ISVLSI)*, pp. 405–410, 2018.
39. Z. U. Abideen, T. D. Perez, and S. Pagliarini, "From FPGAs to obfuscated eASICs: Design and security trade-offs," in *2021 Asian Hardware Oriented Security and Trust Symposium (AsianHOST)*, pp. 1–4, 2021.

40. G. Kolhe, S. M. PD, S. Rafatirad, H. Mahmoodi, A. Sasan, and H. Homayoun, "On custom LUT-based obfuscation," in *Proceedings of the 2019 on Great Lakes Symposium on VLSI*, GLSVLSI '19, p. 477–482, Association for Computing Machinery, 2019.
41. G. Kolhe, H. M. Kamali, M. Naicker, T. D. Sheaves, H. Mahmoodi, P. D. Sai Manoj, H. Homayoun, S. Rafatirad, and A. Sasan, "Security and complexity analysis of LUT-based obfuscation: From blueprint to reality," in *2019 IEEE/ACM International Conference on Computer-Aided Design (ICCAD)*, pp. 1–8, 2019.
42. G. Kolhe, S. Salehi, T. D. Sheaves, H. Homayoun, S. Rafatirad, M. P. Sai, and A. Sasan, "Securing hardware via dynamic obfuscation utilizing reconfigurable interconnect and logic blocks," in *2021 58th ACM/IEEE Design Automation Conference (DAC)*, pp. 229–234, IEEE, 2021.
43. A. Attaran, T. D. Sheaves, P. K. Mugula, and H. Mahmoodi, "Static design of spin transfer torques magnetic look up tables for ASIC designs," in *Proceedings of the 2018 on Great Lakes Symposium on VLSI*, pp. 507–510, 2018.
44. T. Winograd, H. Salmani, H. Mahmoodi, K. Gaj, and H. Homayoun, "Hybrid STT-CMOS designs for reverse-engineering prevention," in *Proceedings of the 53rd Annual Design Automation Conference*, pp. 1–6, 2016.
45. J. Yang, X. Wang, Q. Zhou, Z. Wang, H. Li, Y. Chen, and W. Zhao, "Exploiting spin-orbit torque devices as reconfigurable logic for circuit obfuscation," *IEEE Transactions on Computer-Aided Design of Integrated Circuits and Systems*, vol. 38, no. 1, pp. 57–69, 2018.
46. G. Kolhe, T. D. Sheaves, S. M. P. D., H. Mahmoodi, S. Rafatirad, A. Sasan, and H. Homayoun, "Breaking the design and security trade-off of look-up table-based obfuscation," *ACM Trans. Des. Autom. Electron. Syst.*, 2022.
47. P. Mohan, O. Atli, J. Sweeney, O. Kibar, L. Pileggi, and K. Mai, "Hardware redaction via designer-directed fine-grained eFPGA insertion," in *2021 Design, Automation & Test in Europe Conference & Exhibition (DATE)*, pp. 1186–1191, IEEE, 2021.
48. J. Chen and B. C. Schafer, "Area efficient functional locking through coarse grained runtime reconfigurable architectures," in *Proceedings of the 26th Asia and South Pacific Design Automation Conference*, pp. 542–547, 2021.
49. M. M. Shihab, B. Ramanidharan, S. S. Tellakula, G. Rajavendra Reddy, J. Tian, C. Sechen, and Y. Makris, "ATTEST: Application-agnostic testing of a novel transistor-level programmable fabric," in *2020 IEEE 38th VLSI Test Symposium (VTS)*, pp. 1–6, 2020.
50. J. A. Roy, F. Koushanfar, and I. L. Markov, "EPIC: Ending piracy of integrated circuits," in *2008 Design, Automation and Test in Europe*, pp. 1069–1074, 2008.
51. S. Patnaik, N. Rangarajan, J. Knechtel, O. Sinanoglu, and S. Rakheja, "Advancing hardware security using polymorphic and stochastic spin-hall effect devices," in *2018 Design, Automation & Test in Europe Conference & Exhibition (DATE)*, pp. 97–102, IEEE, 2018.
52. C. M. Tomajoli, L. Collini, J. Bhandari, A. K. T. Moosa, B. Tan, X. Tang, P.-E. Gaillardon, R. Karri, and C. Pilato, "ALICE: An automatic design flow for eFPGA redaction," in *Proceedings of the 59th ACM/IEEE Design Automation Conference*, p. 781–786, 2022.
53. N. Rangarajan, S. Patnaik, J. Knechtel, R. Karri, O. Sinanoglu, and S. Rakheja, "Opening the doors to dynamic camouflaging: Harnessing the power of polymorphic devices," *IEEE Transactions on Emerging Topics in Computing*, 2020.
54. C. Sathe, Y. Makris, and B. C. Schafer, "Investigating the effect of different eFPGAs fabrics on logic locking through HW redaction," in *2022 IEEE 15th Dallas Circuit And System Conference (DCAS)*, pp. 1–6, IEEE, 2022.
55. J. Rajendran, Y. Pino, O. Sinanoglu, and R. Karri, "Security analysis of logic obfuscation," in *DAC Design Automation Conference 2012*, pp. 83–89, 2012.

56. M. Yasin, B. Mazumdar, S. S. Ali, and O. Sinanoglu, "Security analysis of logic encryption against the most effective side-channel attack: DPA," in *2015 IEEE International Symposium on Defect and Fault Tolerance in VLSI and Nanotechnology Systems (DFTS)*, pp. 97–102, 2015.
57. S. Kaveh and J. Yier, "Neos: Netlist encryption and obfuscation suite." https://cadforassurance.org/tools/evaluation-of-obfuscation/neos/, 2024. Accessed: August 16, 2024.
58. S. Roshanisefat, H. Mardani Kamali, H. Homayoun, and A. Sasan, "Rane: An open-source formal de-obfuscation attack for reverse engineering of logic encrypted circuits," GLSVLSI '21, (New York, NY, USA), p. 221–228, Association for Computing Machinery, 2021.
59. R. S. Rajarathnam, Y. Lin, Y. Jin, and D. Z. Pan, "ReGDS: A reverse engineering framework from GDSII to gate-level netlist," in *2020 IEEE International Symposium on Hardware Oriented Security and Trust (HOST)*, pp. 154–163, 2020.
60. W. Che, F. Saqib, and J. Plusquellic, "Puf-based authentication," in *2015 IEEE/ACM International Conference on Computer-Aided Design (ICCAD)*, pp. 337–344, 2015.
61. A. Duncan, F. Rahman, A. Lukefahr, F. Farahmandi, and M. Tehranipoor, "FPGA bitstream security: A day in the life," in *2019 IEEE International Test Conference (ITC)*, pp. 1–10, 2019.
62. Xilinx, Inc., "Using encryption and authentication to secure an ultrascale/ultrascale+ FPGA bitstream," last accessed on Oct 20, 2022. Available at: https://www.xilinx.com/content/dam/xilinx/support/documents/application_notes/xapp1267-encryp-efuse-program.pdf.
63. A. Moradi and T. Schneider, "Improved side-channel analysis attacks on xilinx bitstream encryption of 5, 6, and 7 series," in *Constructive Side-Channel Analysis and Secure Design* (F.-X. Standaert and E. Oswald, eds.), (Cham), pp. 71–87, Springer International Publishing, 2016.
64. F. Benz, A. Seffrin, and S. A. Huss, "Bil: A tool-chain for bitstream reverse-engineering," in *22nd International Conference on Field Programmable Logic and Applications (FPL)*, pp. 735–738, 2012.
65. P. Swierczynski, "Bitstream-based attacks against reconfigurable hardware," last accessed on Oct, 10 2023. Available at: https://www.langer-emv.de/fileadmin/2017_Bitstream-basedattacksagainstreconfigurablehardware-34.pdf.
66. National Institute of Standards and Technology (NIST), "Advanced Encryption Standard (AES)," last accessed on Aug 20, 2023. Available at: https://csrc.nist.gov/files/pubs/fips/197/final/docs/fips-197.pdf.
67. Intrisic ID, "SRAM PUF: The secure silicon fingerprint," last accessed on Sep 11, 2023. Available at: https://www.intrinsic-id.com/wp-content/uploads/2023/03/2023-03-09-White-Paper-SRAM-PUF-The-Secure-Silicon-Fingerprint.pdf.
68. S. Gören, O. Ozkurt, A. Yildiz, and H. F. Ugurdag, "FPGA bitstream protection with PUFs, obfuscation, and multi-boot," in *6th International Workshop on Reconfigurable Communication-Centric Systems-on-Chip (ReCoSoC)*, pp. 1–2, 2011.
69. S. S. Mansouri and E. Dubrova, "Ring oscillator physical unclonable function with multi level supply voltages," in *2012 IEEE 30th International Conference on Computer Design (ICCD)*, pp. 520–521, 2012.
70. K. Fruhashi, M. Shiozaki, A. Fukushima, T. Murayama, and T. Fujino, "The arbiter-PUF with high uniqueness utilizing novel arbiter circuit with delay-time measurement," in *2011 IEEE International Symposium of Circuits and Systems (ISCAS)*, pp. 2325–2328, 2011.
71. J. Miskelly and M. O' Neill, "Fast DRAM PUFs on commodity devices," *IEEE Transactions on Computer-Aided Design of Integrated Circuits and Systems*, vol. 39, no. 11, pp. 3566–3576, 2020.
72. S. Zhang, B. Gao, D. Wu, H. Wu, and H. Qian, "Evaluation and optimization of physical unclonable function (PUF) based on the variability of FinFET SRAM," in *2017 International Conference on Electron Devices and Solid-State Circuits (EDSSC)*, pp. 1–2, 2017.

73. K.-H. Chuang, E. Bury, R. Degraeve, B. Kaczer, D. Linten, and I. Verbauwhede, "A physically unclonable function using soft oxide breakdown featuring 0% native BER and 51.8 fj/bit in 40-nm CMOS," *IEEE Journal of Solid-State Circuits*, vol. 54, no. 10, pp. 2765–2776, 2019.
74. R. Maes, V. Rozic, I. Verbauwhede, P. Koeberl, E. van der Sluis, and V. van der Leest, "Experimental evaluation of physically unclonable functions in 65 nm CMOS," in *2012 Proceedings of the ESSCIRC (ESSCIRC)*, pp. 486–489, 2012.
75. S. Baek, G.-H. Yu, J. Kim, C. T. Ngo, J. K. Eshraghian, and J.-P. Hong, "A reconfigurable SRAM based CMOS PUF with challenge to response pairs," *IEEE Access*, vol. 9, pp. 79947–79960, 2021.
76. Y. Shifman, A. Miller, Y. Weizmann, and J. Shor, "A 2 bit/cell tilting sram-based PUF with a BER of 3.1e-10 and an energy of 21 fj/bit in 65nm," *IEEE Open Journal of Circuits and Systems*, vol. 1, pp. 205–217, 2020.
77. A. B. Alvarez, W. Zhao, and M. Alioto, "Static physically unclonable functions for secure chip identification with 1.9–5.8% native bit instability at 0.6–1 v and 15 fj/bit in 65 nm," *IEEE Journal of Solid-State Circuits*, vol. 51, no. 3, pp. 763–775, 2016.
78. G.-J. Schrijen and V. van der Leest, "Comparative analysis of SRAM memories used as PUF primitives," in *2012 Design, Automation & Test in Europe Conference & Exhibition (DATE)*, pp. 1319–1324, 2012.
79. G. Selimis, M. Konijnenburg, M. Ashouei, J. Huisken, H. de Groot, V. van der Leest, G.-J. Schrijen, M. van Hulst, and P. Tuyls, "Evaluation of 90nm 6t-sram as physical unclonable function for secure key generation in wireless sensor nodes," in *2011 IEEE International Symposium of Circuits and Systems (ISCAS)*, pp. 567–570, 2011.
80. R. Wang, G. Selimis, R. Maes, and S. Goossens, "Long-term continuous assessment of SRAM PUF and source of random numbers," in *2020 Design, Automation & Test in Europe Conference & Exhibition (DATE)*, pp. 7–12, 2020.
81. M. Neve and K. Tiri, "On the complexity of side-channel attacks on AES-256 – methodology and quantitative results on cache attacks." Cryptology ePrint Archive, Paper 2007/318, 2007. https://eprint.iacr.org/2007/318.
82. H. Ma, S. Pan, Y. Gao, J. He, Y. Zhao, and Y. Jin, "Vulnerable pqc against side channel analysis - a case study on kyber," in *2022 Asian Hardware Oriented Security and Trust Symposium (AsianHOST)*, pp. 1–6, 2022.
83. A. Moradi, "Side-channel leakage through static power," in *Cryptographic Hardware and Embedded Systems – CHES 2014* (L. Batina and M. Robshaw, eds.), (Berlin, Heidelberg), pp. 562–579, Springer Berlin Heidelberg, 2014.
84. N. Gattu, M. N. Imtiaz Khan, A. De, and S. Ghosh, "Power side channel attack analysis and detection," in *2020 IEEE/ACM International Conference On Computer Aided Design (ICCAD)*, pp. 1–7, 2020.
85. G. Kolhe, T. Sheaves, K. I. Gubbi, S. Salehi, S. Rafatirad, S. M. PD, A. Sasan, and H. Homayoun, "Lock&roll: deep-learning power side-channel attack mitigation using emerging reconfigurable devices and logic locking," in *Proceedings of the 59th ACM/IEEE Design Automation Conference*, pp. 85–90, 2022.

Classification of ReBO Techniques 4

4.1 Classification of ReBO Approaches

FPGA devices primarily inspire the ReBO concept, although it has not garnered as much attention as LL in research. Notably, a pivotal study by authors from Microsoft and Iowa State University, detailed in [1], effectively recognized the potential of reconfigurable logic as an obfuscation resource. This initial work has since paved the way for further research endeavors to enhance digital design security through ReBO techniques. However, the interest in utilizing reconfigurable logic for obfuscation techniques grows within the research community, and it is essential to establish a standardized classification and terminology for ReBO.

The ReBO can be implemented using either coarse-grain or fine-grain reconfigurable elements. In both cases, the key element is the LUT, which provides reconfigurability by requiring configuration bits to define the boolean function. These configuration bits need to be stored in memory, which can be accomplished through various technologies. Apart from LUTs, switch boxes can also be used to achieve obfuscation. ReBO can be implemented at different levels of IPs. This classification of ReBO is based on three main factors: the technology used, the type of element, and the type of IP. An overview of this classification can be seen in Fig. 4.1.

4.1.1 Technology

Numerous obfuscation techniques based on reconfigurability have been proposed, with LUTs playing a crucial role in enabling logic obfuscation [4, 5]. During the obfuscation process, specific internal gates from the design are assigned to LUTs. Figure 4.1 illustrates the various technologies available for storing the essential bits of these LUTs, which can

Z. U. Abideen and S. Pagliarini, *Reconfigurable Obfuscation Techniques for the IC Supply Chain*, Synthesis Lectures on Digital Circuits & Systems,
https://doi.org/10.1007/978-3-031-77509-3_4

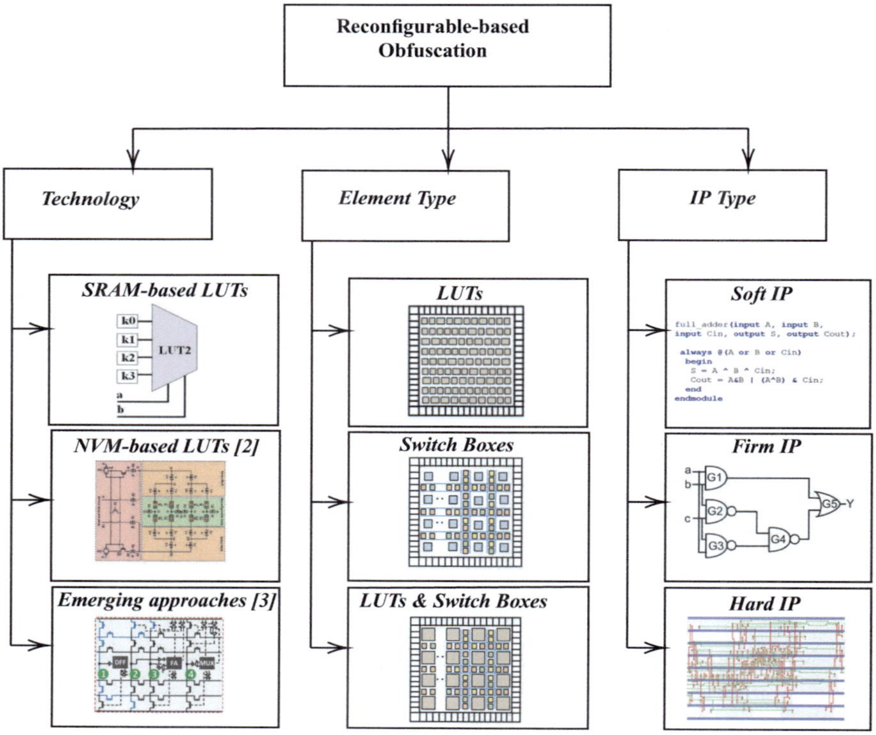

Fig. 4.1 Classification of ReBO by technology, element type, and IP category

be classified into three distinct groups. These groups include SRAM-based LUTs [1, 6, 7], NVM-based LUTs [8–10], and emerging approaches [3, 11, 12]. While SRAM-based LUTs have been widely used for programming LUTs, NVM-based LUTs have gained popularity due to their enhanced security features [13, 14]. The category of emerging approaches encompasses various methods for programming the LUTs, including FF-based LUTs and individual transistor programming, such as in TRAP fabric and the programming of efuses.

The SRAM-based LUT is a popular technology for storing key bits due to its desirable characteristics, including programmability, reconfigurability, fast access time, low power consumption, small area, scalability, and ease of testing. These features make it an ideal choice for implementing logic functions in FPGA designs, and they are also suitable for obfuscation purposes. A 2-input LUT is shown in Fig. 4.2, demonstrating the various functions it can potentially implement. With two inputs, it can realize 16 distinct functions, as detailed in Fig. 4.2.

NVM-based LUTs make use of non-volatile memory technologies such as MRAM, which is designed with STT-MTJ or STT-MTJ devices. The NVM-based LUTs exploit the aforementioned emerging technologies, which can be categorized as STT-MTJ-based LUT and

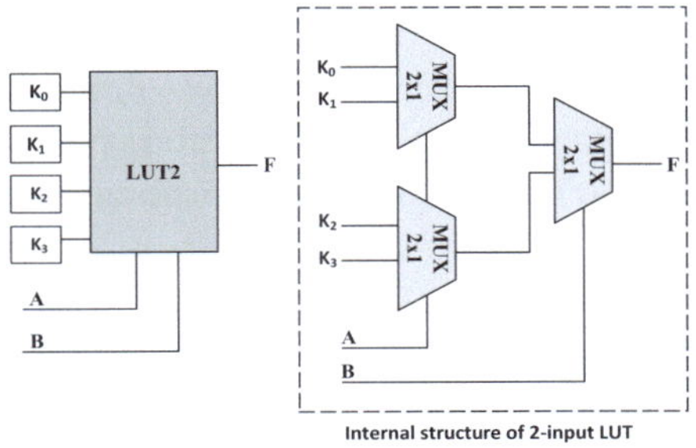

K_3	K_2	K_1	K_0	F	Gate
0	0	0	0	0	TIE0
0	0	0	1	A'B'	NOR
0	0	1	0	AB'	INR
0	0	1	1	B'	INV
0	1	0	0	A'B	INR
0	1	0	1	A'	INV
0	1	1	0	AB'+A'B	XOR
0	1	1	1	A'+B'	NAND

K_3	K_2	K_1	K_0	F	Gate
1	0	0	0	AB	AND
1	0	0	1	A'B'+AB	XNOR
1	0	1	0	A	BUF
1	0	1	1	A+B'	IND
1	1	0	0	B	BUF
1	1	0	1	A'+B	IND
1	1	1	0	A+B	OR
1	1	1	1	1	TIE1

Fig. 4.2 The internal architecture of the 2-input LUT and all the possible logic functions based on its configuration bits K_0-K_3

SOT-MTJ-based LUT. To understand the concept, let us consider the operation of an STT-MTJ-based LUT. In this case, the state switching of STT-MTJ requires a dual-directional current source. As a result, four NMOS (MN0 through MN3) powered by Vdda and Gnda are used to produce the current passing through the STT-MTJs as shown in Fig. 4.3. Only two NMOS are active at any given time, controlled by a circuit composed of two NOR logic gates. These gates define the activation based on the control signal EN and the direction of current based on input. Figure 4.3 illustrates a 2-input STT-MTJ-based LUT with four STT-MTJs. The reconfiguration of this LUT can be carried out bit by bit in a series to optimize the chip area. In this scenario, only one current source is necessary, but each STT-MTJ should be associated with an additional NMOS transistor, providing it with corresponding addresses (e.g., $C0 - C3$). The STT effect has a very fast switching speed of less than 1ns [15]. The series reconfiguration maintains high speed and can reconfigure complex STT-MTJ-based LUTs with over five inputs in just a few hundred nanoseconds.

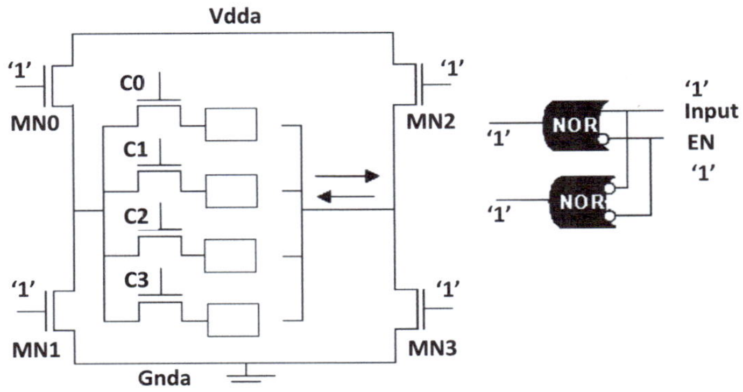

Fig. 4.3 An example of 2-inputs STT-MTJ-based LUT [16]

NVM-based LUT maintains programming even when the device is not powered. However, compared to SRAM-based LUTs, NVM-based LUTs have slower access times and more limited and complex programmability. On the other hand, NVM-based LUTs provide high-density storage elements, and STT-MTJ-based LUTs show promise for creating robust and resistant LUTs.

The use of FF-based LUT implementation allows for technology-agnostic framework, simplifying the floorplanning and placement processes. However, it is important to note that FF-based LUTs do not achieve the same bit density level as SRAM-based solutions. The TRAP fabric is a unique approach for concealing the design's intent by programming numerous transistors. On the other hand, efuses are one-time programmable fuses permanently programmed with a specific LUT configuration. This differs from SRAM-based LUTs, which require reloading each time the device powers up. While efuses offer distinct security properties by preventing reprogramming, they may also expose programmed values to reverse engineering by end-users (but not by the foundry).

4.1.1.1 Element Type

When categorizing ReBO, one can classify them based on the type of components they use. In this approach, the LUT acts as a locking element for providing obfuscation. Some approaches solely rely on LUTs for obfuscation purposes, as various studies have demonstrated [1, 6, 7, 13, 14, 17]. These methods mainly focus on obfuscating the logic elements by using LUTs alone. In this type of approach, the LUTs are designed with the technologies discussed in the previous section [4, 9, 10]. The targeted obfuscation logic is mapped to the LUTs with the given obfuscation criteria. Researchers have also proposed various selection methods where they select the logic based on various selection methods, such as output corruptibility, data path, and functional traces [14]. This gives the designer the freedom to design the n-input LUT. The LUTs with different input sizes could be targeted for different parts of the design. For example, the part of the design that is less sensitive to security could be targeted with LUTs having a small number of inputs. These decisions are part of the selection algorithms. It has been demonstrated that circuits with a small LUT input size, such as a 2-input LUT, are prone to easy de-obfuscation [14].

When it comes to implementation, these LUTs are integrated with standard cells, offering options in both custom and digital implementation methodologies. Custom implementation can achieve a highly compact area. However, this approach might come with limitations, particularly in terms of flexibility. The fixed nature of custom implementation could restrict the ability to vary the input size of the LUT, potentially constraining the design's adaptability to different security requirements.

Digital implementation offers more flexibility, allowing for easier adjustments to the LUT's input size. This flexibility can be crucial in applications where the design needs to accommodate varying configurations or adapt to evolving standards. However, the trade-off is often in the area efficiency, as digital implementations may not achieve the same level of compactness as custom approaches.

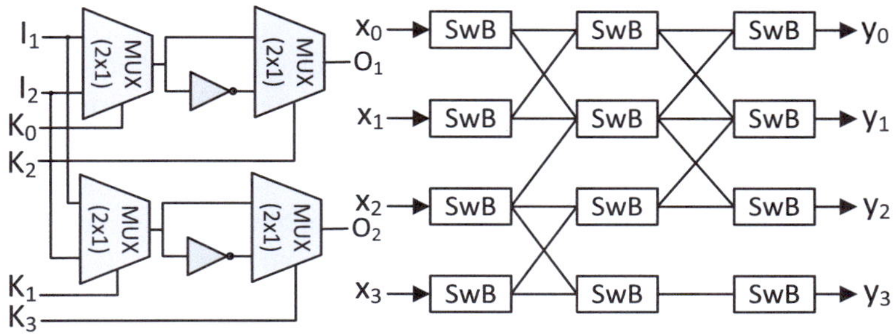

Fig. 4.4 **a** The key-driven switch-box circuit that exchanges the routing of two signals, and **b** one potential configuration of the switching network that creates the SAT-hard instance [19]

On the other hand, there are obfuscation techniques that exclusively rely on exploiting switch boxes [3, 18, 19]. These methods alter the connections between elements to obfuscate the design, as shown in the center of Fig. 4.1. In this case, the layout combines switch boxes and standard logic, with the switch boxes highlighted in blue. Figure 4.4 displays the construction of the switch-boxes present as a sample network topology and the placement of secure instances in the netlist by building a switching network using Muxes. In various switching networks, there is a series of self-routing logarithmic networks, known as $\log_2 N$ networks, which offer customizable interconnection with reduced overhead when compared to traditional networks like mesh or crossbar. Each switching block allows the outputs to be arranged in any permutation of the inputs. Additionally, the incorporated key-configurable inverters are included in each wire, enabling the output to be rearranged and inverted based on the key value.

Lastly, some techniques use a combination of both LUTs and switch boxes. These are generally known as FPGA redaction techniques [20–23]. The architecture of the eFPGA is similar to that of an FPGA, incorporating not only LUTs and Muxes, but also FFs [24, 25]. The targeted part of the design is implemented on the eFPGA. The architecture of the eFPGA could be standard for commercial FPGA sellers, or the designer could also customize the architecture. This opportunity is available with open-source tools, where the designer specifies and the tool generates the architecture [26]. After the obfuscation, the design needs implementation, which requires significant effort from the designer from the placement and routing point of view.

As research on this technique has progressed, the trade-off between PPA versus security has become increasingly important. It has been found that the size of the LUT input plays a crucial and direct role in achieving resilience against attacks. In the case of eFPGA redaction, the design focuses on minimizing the size of the eFPGA tile. This technique also requires the development of a custom tool that should be automated and compatible with the traditional CAD flow.

4.1.1.2 IP Type

When implementing ReBO techniques, designers need to follow additional steps in the design process. They must choose crucial modules to be redacted or identify appropriate gates to replace using LUTs. The obfuscation process can occur at various IP levels, namely soft IP, firm IP, or hard IP, as shown in Fig. 4.1. Soft IP obfuscation involves altering the RTL code, such as Verilog or VHDL, or high-level codes like C/C++, through user-defined algorithms to identify and obfuscate specific parts of the modules [11, 21, 27, 28].

When using Verilog or VHDL for obfuscation, the custom tool of the corresponding category reads the original design and divides it into reconfigurable and non-reconfigurable parts. Next, the reconfigurable part is translated into logic on the HDL level [20, 23, 29]. All of this takes place before logic synthesis. It is important to note that the obfuscated design, including the reconfigurable and non-reconfigurable parts, is subjected to various design obfuscation techniques, such as clock gating [30].

Regarding the same category, if we move another level to the HLS level, then the framework reads the design description in C/C++ and partitions the design. Then, it performs the HLS synthesis [27, 29]. In the process of synthesis, the synthesizer arranges various operations in the code by taking into account data dependencies and the delay information of each operator. To ensure successful operation scheduling, the synthesizer must have knowledge of the delay for each functional unit [27]. Commercial HLS tools offer library characterizers that extract area and delay information for every basic primitive, tailored to the target technology.

Also, challenges arise with function encapsulation and pragma annotation. Function encapsulation controls parallelism by managing the number of hardware instances, which requires precise constraint settings to avoid inefficiency. Pragma annotation involves identifying and marking lines of code that can be isolated and mapped as functional operators, a task complicated by complex syntax like loops and conditionals [29]. To mitigate these issues, incremental testing, modular design, and automated tools can be used to fine-tune the synthesis process, ensuring efficient hardware utilization and accurate code mapping.

The most common approach to firm IP obfuscation uses the post-synthesis netlist file as input and generates the obfuscated netlist as output [1, 3, 5–8, 10, 20]. This level of obfuscation contributes best to automation. The obfuscation tool can compute the initial PPA overheads and make trade-offs when selecting the gate to be mapped to the reconfigurable part. Considering the security versus design trade-offs, this level of obfuscation will consume less runtime and provide a good estimate of the PPA overheads. For example, when selecting the gate for the LUT's candidate, one of the parameters could be fan-in/fan-out, highlighting the benefits of obfuscation at the firm IP level [13, 14].

The process of hard IP obfuscation involves identifying a critical part of the design and integrating it into a pre-designed and pre-verified block of logic known as hard IP. This is typically done using eFPGAs for obfuscation purposes [22, 25]. These hard IP blocks have fixed physical layouts and are directly integrated into the chip's floorplan during the design phase, which requires careful consideration of placement, pin alignment, and routing.

This approach is more suitable for eFPGA redaction when the reconfigurable part is coarse-grained and not spread around the design. Since hard IP blocks have fixed dimensions and timing characteristics, they require precise timing closure and power distribution planning. Unlike soft IP, which is more flexible, hard IP is optimized for specific process nodes and offers limited customization. Therefore, it is essential for designers to carefully select and configure these blocks. Once integrated, hard IP blocks can be reused across multiple designs.

4.2 Takeaway Notes

The security of ReBO relies heavily on the reconfigurable part that undergoes obfuscation [27]. LUTs based on SRAM are compact and dense, but using thousands of them leads to overhead. NVM-based LUTs have lower PPA overheads compared to SRAM-based LUTs. For example, the authors in [4] achieved low PPA overheads for small circuits or benchmarks. However, for a high obfuscation rate, the average area overhead is 262 times higher than the baseline designs. In general, both LUT-based obfuscation with SRAM technology and eFPGA redaction techniques have high PPA overheads. Conversely, NVM-based LUTs with emerging technologies introduces additional challenges in fabrication.

So far, we have discussed the classification of ReBO techniques. In Chap. 5, we will describe recent attacks specifically targeting ReBO techniques.

References

1. A. Baumgarten, A. Tyagi, and J. Zambreno, "Preventing IC piracy using reconfigurable logic barriers," *IEEE Design Test of Computers*, vol. 27, no. 1, pp. 66–75, 2010.
2. G. Kolhe, T. Sheaves, K. I. Gubbi, S. Salehi, S. Rafatirad, S. M. PD, A. Sasan, and H. Homayoun, "Lock&roll: deep-learning power side-channel attack mitigation using emerging reconfigurable devices and logic locking," in *Proceedings of the 59th ACM/IEEE Design Automation Conference*, pp. 85–90, 2022.
3. M. M. Shihab, B. Ramanidharan, S. S. Tellakula, G. Rajavendra Reddy, J. Tian, C. Sechen, and Y. Makris, "ATTEST: Application-agnostic testing of a novel transistor-level programmable fabric," in *2020 IEEE 38th VLSI Test Symposium (VTS)*, pp. 1–6, 2020.
4. G. Kolhe, T. Sheaves, K. I. Gubbi, T. Kadale, S. Rafatirad, S. M. PD, A. Sasan, H. Mahmoodi, and H. Homayoun, "Silicon validation of LUT-based logic-locked IP cores," in *Proceedings of the 59th ACM/IEEE Design Automation Conference*, pp. 1189–1194, 2022.
5. S. D. Chowdhury, G. Zhang, Y. Hu, and P. Nuzzo, "Enhancing SAT-attack resiliency and cost-effectiveness of reconfigurable-logic-based circuit obfuscation," in *2021 IEEE International Symposium on Circuits and Systems (ISCAS)*, pp. 1–5, IEEE, 2021.
6. B. Liu and B. Wang, "Embedded reconfigurable logic for ASIC design obfuscation against supply chain attacks," in *2014 Design, Automation Test in Europe Conference Exhibition (DATE)*, pp. 1–6, 2014.

7. H. Mardani Kamali, K. Zamiri Azar, K. Gaj, H. Homayoun, and A. Sasan, "LUT-lock: A novel LUT-based logic obfuscation for FPGA-bitstream and ASIC-hardware protection," in *2018 IEEE Computer Society Annual Symposium on VLSI (ISVLSI)*, pp. 405–410, 2018.
8. A. Attaran, T. D. Sheaves, P. K. Mugula, and H. Mahmoodi, "Static design of spin transfer torques magnetic look up tables for ASIC designs," in *Proceedings of the 2018 on Great Lakes Symposium on VLSI*, pp. 507–510, 2018.
9. T. Winograd, H. Salmani, H. Mahmoodi, K. Gaj, and H. Homayoun, "Hybrid STT-CMOS designs for reverse-engineering prevention," in *Proceedings of the 53rd Annual Design Automation Conference*, pp. 1–6, 2016.
10. J. Yang, X. Wang, Q. Zhou, Z. Wang, H. Li, Y. Chen, and W. Zhao, "Exploiting spin-orbit torque devices as reconfigurable logic for circuit obfuscation," *IEEE Transactions on Computer-Aided Design of Integrated Circuits and Systems*, vol. 38, no. 1, pp. 57–69, 2018.
11. C. Sathe, Y. Makris, and B. C. Schafer, "Investigating the effect of different eFPGAs fabrics on logic locking through HW redaction," in *2022 IEEE 15th Dallas Circuit And System Conference (DCAS)*, pp. 1–6, IEEE, 2022.
12. Z. U. Abideen, S. Gokulanathan, M. J. Aljafar, and S. Pagliarini, "An overview of FPGA-inspired obfuscation techniques," *Association for Computing Machinery*, vol. 56, no. 12, December 2024.
13. G. Kolhe, S. M. PD, S. Rafatirad, H. Mahmoodi, A. Sasan, and H. Homayoun, "On custom LUT-based obfuscation," in *Proceedings of the 2019 on Great Lakes Symposium on VLSI*, GLSVLSI '19, p. 477–482, Association for Computing Machinery, 2019.
14. G. Kolhe, H. M. Kamali, M. Naicker, T. D. Sheaves, H. Mahmoodi, P. D. Sai Manoj, H. Homayoun, S. Rafatirad, and A. Sasan, "Security and complexity analysis of LUT-based obfuscation: From blueprint to reality," in *2019 IEEE/ACM International Conference on Computer-Aided Design (ICCAD)*, pp. 1–8, 2019.
15. T. Devolder, A. Tulapurkar, K. Yagami, P. Crozat, C. Chappert, A. Fukushima, and Y. Suzuki, "Ultra-fast magnetization reversal in magnetic nano-pillars by spin-polarized current," *Journal of Magnetism and Magnetic Materials*, vol. 286, pp. 77–82, 2005. Proceedings of the 5th International Symposium on Metallic Multilayers.
16. W. Zhao, E. Belhaire, C. Chappert, and P. Mazoyer, "Spin transfer torque (stt)-mram–based runtime reconfiguration fpga circuit," *ACM Trans. Embed. Comput. Syst.*, vol. 9, oct 2009.
17. G. Kolhe, T. D. Sheaves, S. M. P. D., H. Mahmoodi, S. Rafatirad, A. Sasan, and H. Homayoun, "Breaking the design and security trade-off of look-up table-based obfuscation," *ACM Trans. Des. Autom. Electron. Syst.*, 2022.
18. H. Chakraborty and R. Vemuri, "Rtl interconnect obfuscation by polymorphic switch boxes for secure hardware generation," in *2024 25th International Symposium on Quality Electronic Design (ISQED)*, pp. 1–8, 2024.
19. I. McDaniel, M. Zuzak, and A. Srivastava, "Removal of sat-hard instances in logic obfuscation through inference of functionality," vol. 29, jul 2024.
20. P. Mohan, O. Atli, J. Sweeney, O. Kibar, L. Pileggi, and K. Mai, "Hardware redaction via designer-directed fine-grained eFPGA insertion," in *2021 Design, Automation & Test in Europe Conference & Exhibition (DATE)*, pp. 1186–1191, IEEE, 2021.
21. S. Patnaik, N. Rangarajan, J. Knechtel, O. Sinanoglu, and S. Rakheja, "Advancing hardware security using polymorphic and stochastic spin-hall effect devices," in *2018 Design, Automation & Test in Europe Conference & Exhibition (DATE)*, pp. 97–102, IEEE, 2018.
22. C. M. Tomajoli, L. Collini, J. Bhandari, A. K. T. Moosa, B. Tan, X. Tang, P.-E. Gaillardon, R. Karri, and C. Pilato, "ALICE: An automatic design flow for eFPGA redaction," in *Proceedings of the 59th ACM/IEEE Design Automation Conference*, p. 781–786, 2022.

23. N. Rangarajan, S. Patnaik, J. Knechtel, R. Karri, O. Sinanoglu, and S. Rakheja, "Opening the doors to dynamic camouflaging: Harnessing the power of polymorphic devices," *IEEE Transactions on Emerging Topics in Computing*, 2020.
24. M. M. Shihab, J. Tian, G. R. Reddy, B. Hu, W. Swartz, B. Carrion Schaefer, C. Sechen, and Y. Makris, "Design obfuscation through selective post-fabrication transistor-level programming," in *2019 Design, Automation & Test in Europe Conference & Exhibition (DATE)*, pp. 528–533, 2019.
25. J. Bhandari, A. K. Thalakkattu Moosa, B. Tan, C. Pilato, G. Gore, X. Tang, S. Temple, P.-E. Gaillardon, and R. Karri, "Exploring eFPGA-based redaction for IP protection," in *2021 IEEE/ACM International Conference On Computer Aided Design (ICCAD)*, pp. 1–9, 2021.
26. K. E. Murray, O. Petelin, S. Zhong, J. M. Wang, M. Eldafrawy, J.-P. Legault, E. Sha, A. G. Graham, J. Wu, M. J. P. Walker, H. Zeng, P. Patros, J. Luu, K. B. Kent, and V. Betz, "VTR 8: High-performance cad and customizable FPGA architecture modelling," *ACM Transactions on Reconfigurable Technology and Systems*, vol. 13, no. 2, 2020.
27. B. Hu, T. Jingxiang, S. Mustafa, R. R. Gaurav, S. William, M. Yiorgos, C. S. Benjamin, and S. Carl, "Functional obfuscation of hardware accelerators through selective partial design extraction onto an embedded FPGA," in *Proceedings of the 2019 Great Lakes Symposium on VLSI*, p. 171–176, 2019.
28. J. Chen and B. C. Schafer, "Area efficient functional locking through coarse grained runtime reconfigurable architectures," in *Proceedings of the 26th Asia and South Pacific Design Automation Conference*, pp. 542–547, 2021.
29. J. Chen, M. Zaman, Y. Makris, R. D. S. Blanton, S. Mitra, and B. C. Schafer, "DECOY: Deflection-Driven HLS-Based Computation Partitioning for Obfuscating Intellectual PropertY," in *Proceedings of the 57th ACM/EDAC/IEEE Design Automation Conference*, DAC '20, IEEE Press, 2020.
30. G. Basiashvili, Z. U. Abideen, and S. Pagliarini, "Obfuscating the hierarchy of a digital ip," in *Embedded Computer Systems: Architectures, Modeling, and Simulation* (A. Orailoglu, M. Reichenbach, and M. Jung, eds.), (Cham), pp. 303–314, Springer International Publishing, 2022.

Evaluating ReBO: Attack Strategies and Security Analysis

5

5.1 Emerging Attacks

When it comes to adversarial modeling, there are two types of threat models: oracle-guided and oracle-less. In an oracle-guided attack, the adversary possesses reverse-engineered locked netlist and a functional IC, which acts as an oracle. Oracle-guided SAT attacks and their variations can be executed on ReBO techniques, which are quite common [1–5]. On the other hand, oracle-less attacks do not require access to an oracle or functional IC; instead, they only need the targeted design for de-obfuscation. Despite the efforts, these attacks are unsuccessful in finding the bitstream of ReBO [6].

To gain a better understanding of the attacks on ReBO, we will first explain the oracle-guided attack, which is best illustrated by the SAT attack. Then, we will explain the existing oracle-less attacks that were directed at LL but can also be applied to ReBO. Finally, we will discuss the emerging and innovative attacks that were specifically aimed at compromising the security of ReBO. It is noteworthy that, so far, *none of these attacks completely broke ReBO's security*. The details of these attacks are provided in the subsequent subsections.

5.1.1 Oracle-Guided Attacks

Let us understand the basic working principle of the SAT attack to recover a key. SAT attack starts by creating two versions of the locked circuit, namely L_{A} and L_{B}, by using different keys, K_{A} and K_{B} respectively. Then, it generates the miter circuit, which checks if there is a difference between the outputs of the two circuits, as shown in Fig. 5.1. In this figure, the primary inputs (I) are shared between the two locked circuits and the $diff$ output is generated by XORing the corresponding outputs of the two circuits and then ORing them.

Z. U. Abideen and S. Pagliarini, *Reconfigurable Obfuscation Techniques for the IC Supply Chain*, Synthesis Lectures on Digital Circuits & Systems,
https://doi.org/10.1007/978-3-031-77509-3_5

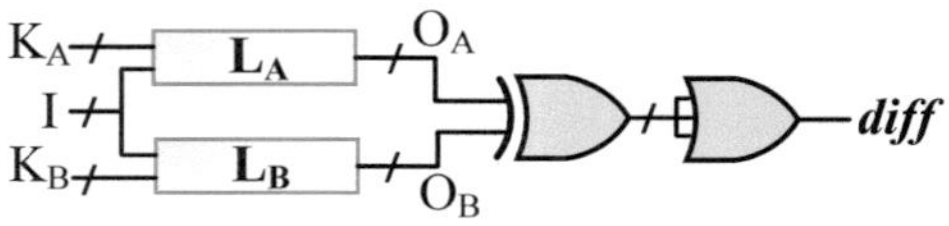

Fig. 5.1 Circuit employed by the SAT attack

Algorithm 1 describes the SAT attack, which iteratively finds the distinguishing input pattern (DIP) to eliminate incorrect keys.

The SAT attack begins by identifying the conjunctive normal form (CNF) formulas of L_A and L_B, before generating the function F_1 on line 2. It then enters a loop on lines 3–7, whenever there is a satisfiable solution on the conjunction of F_i with the CNF formula of the miter circuit. This satisfiable solution means that there exists a DIP I_d, which causes circuits L_A and L_B to generate incorrect outputs. The SAT assignment on line 4 is used to extract this DIP, which is then applied to the oracle (R) to obtain the output O_d. The CNF formula F_i is updated with the additional information obtained when O_d and I_d are applied to circuits L_A and L_B on line 6. The loop continues until no more DIPs are found. The correct key K_C is found as an assignment to K_A that satisfies the formula F_i on line 8. Note that while the SAT attack and its variants are powerful techniques, they may face issues with circuits that are locked by a large number of key bits, such as reconfigurable-based obfuscation techniques.

Previously, we described the basic SAT attack that was initially developed by Subramanyan [1]. After that, a series of LL attacks and defenses came about, a process which is still ongoing like a cat and mouse game these days [1, 3–5, 7–9]. These variations make use of traditional SAT attacks and, at times, an additional security loophole is exploited. The attack in [2] demonstrates that an attacker can decipher the obfuscated netlist in a time linear to the number of keys by sensitizing the key values to the output. We then develop techniques to fix this vulnerability and make obfuscation truly exponential in the number of inserted keys.

Algorithm 1: SAT attack [1]

Input : Locked netlist L, Oracle R
Output: Correct key K_C

1 $i \leftarrow 1$;
2 $F_1 \leftarrow L(I, K_A, O_A) \wedge L(I, K_A, O_B)$;
3 **while** $SAT[F_i \wedge (O_A \neq O_B)]$ **do**
4 $\quad I_d \leftarrow \text{SAT\_ASSIGNMENT}_I[F_i \wedge (O_A \neq O_B)]$;
5 $\quad O_d \leftarrow R(I_d)$;
6 $\quad F_{i+1} \leftarrow F_i \wedge L(I_d, K_A, O_d) \wedge L(I_d, K_B, O_d)$;
7 $\quad i \leftarrow i + 1$;
8 $K_C \leftarrow \text{SAT\_ASSIGNMENT}_{K_A}(F_i)$;
9 **return** K_C;

The attack in [7] is a SAT-based sensitization attack consisting of two main stages: the sensitization and the SAT execution per se. The sensitization stage calculates feasible attack patterns used to guide the SAT attack. The SAT attack uses the patterns from the sensitization stage to extract the correct key efficiently. The sensitization stage also helps identify dummy AND/OR trees and prevents the SAT attack from encountering long trails. This attack also utilizes input bias to distinguish between dummy and real trees, reducing the locked netlist to ensure a successful SAT attack. The correctness of the retrieved key is validated by applying the computed pattern to the functional IC, i.e., the oracle.

The AppSAT attack aims to reduce a multi-layered defense to a single layer by querying the functional IC with a fixed number of random DIPs at regular intervals and augmenting the CNF formula with new constraints based on these DIPs [4]. The attack terminates when the Hamming distance between the correct output from the functional IC and the locked netlist is very low and returns an approximately correct key that yields an approximate netlist. This approximate key can be used as a pre-processing attack to peel off defenses one at a time.

The CycSAT attack leverages the cycles in SAT-based attacks on circuits [3]. The first one assumes that at least one correct key will reveal a clear path in the system and adds a rule to prevent confusing loops. The second approach assumes that a correct key will reveal a straightforward but possibly confusing path. Before using the original SAT attack, a formula is included to suggest that there are no detectable loops in the circuit. Additional attempts are required to eliminate hidden recurring patterns and cycles after the attack has been carried out.

5.1.2 Oracle-Less Attacks

When looking at oracle-less attacks, we can see that these types of attacks do not require functional IC but instead assume that the adversary has only the locked netlist. Structural analysis serves as the basis for several oracle-less attacks, which are discussed in [2, 10–15]. For instance, the attack detailed in [10] is carried out through a divide-and-conquer approach. This method involves dividing a circuit into logic cones and then targeting individual logic cones using brute force. A logic cone is a sub-circuit composed of gates within the transitive fan-in of a specific primary output. Constructing a logic cone, from the primary output back to the primary input(s), is equivalent to a depth-first search in a graph.

The attack in [11] leverages the structural traces in AntiSAT block [16] to identify and isolate the AntiSAT block. The AntiSAT block is created so that the total number of SAT attack iterations, and consequently the overall execution time required to obtain the correct key, increases exponentially with the key size in the AntiSAT block. The authors found that inserting key gates into the AntiSAT block affects the behavior of the gates. After adding the key gates for obfuscation, the skew of the targeted gate changes depending on the placement of the key gates. For example, if we consider an n-input AND gate in the AntiSAT block, inserting an XOR key gate at a certain point changes the skew of the AND gate. The location of the key gate determines how much the skew changes.

The attack described in [12] uses structural traces and machine learning (ML) for vulnerability analysis. The attack makes observations about implementing structural alterations in traditional LL. Motivated by these observations, the attack introduces a structural analysis using ML, which exposes critical vulnerabilities in LL approaches. It extracts each LL gate locality from a locked design and reverts them to their pre-synthesis state using reconstruction models trained on locality pairs. This information allows attackers to conduct key guessing and reverse engineering attacks more easily.

The constant propagation attack on LL presented in [13] exploits changes that occur during the synthesis process. This attack conducts synthesis analysis on each key input by optimizing the locked system with hard-coded correct and incorrect values for the key bit. It is scalable and does not necessitate an unlocked circuit, although having an unlocked system is advised for functional tests and to verify the correctness of the extracted key. The proposed approach analyzes information provided by the synthesis process in a generated report. This attack is characterized by precision, reducing brute force efforts to identify the key. It is challenging to prevent due to the naivety and predictable patterns of current LL techniques. Unlike many existing attacks, this attack exploits the structural effects of the locked circuit rather than performing functional analysis, exposing more vulnerabilities.

The proposed attack, SCOPE [14], is suitable for various design types, even those that may be intricate for SAT attacks and highly scalable to any circuit size. SCOPE draws inspiration from [13], a training-based constant propagation attack as given in [13], which analyzes the design's structure to recover the correct key value. While its core concept shares similarities with the attack given in [13], SCOPE is augmented to operate unsupervised, eliminating the need for training data while achieving comparable or superior performance. The emphasis on unsupervised analysis serves two primary objectives: firstly, to obviate the necessity for the attacker to possess knowledge about the applied locking algorithm, and secondly, to enhance scalability and reduce attack runtime. SCOPE represents an unsupervised constant propagation attack on LL, leveraging existing synthesis/optimization tools to scrutinize each key input port for critical features that may disclose the correct key value.

5.2 Specialized ReBO Attacks

Recent attacks on ReBO have been presented in [6, 17, 18]. In the next subsection, we will describe in detail these attacks.

5.2.1 Predictive Model Attack

This attack replaces the precise logic implemented on eFPGAs with a synthesizable predictive model. This oracle-guided attack uses machine learning techniques to construct a predictive model that aims to replicate the behavior of the original logic [17]. Let us discuss the threat model of the predictive model attack. In this scenario, the adversary is a skilled IC designer with the necessary knowledge and tools to understand the layout representation. The threat model presented in [17] is outlined below:

- An eFPGA's input and output can be accessed by an adversary through the scan-chain that encircles it. FPGA companies commonly utilize this technique in their commercial IPs.
- When a scan-chain is absent, an adversary may opt for a probing attack as a viable alternative [19]. Due to the predictable configuration of the eFPGA, a probing attack can be executed with relative ease.
- Given that the eFPGA is typically obtained through licensed companies, such as Achronix, Menta, or Quicklogic, it is reasonable to assume that the adversary can access the CAD tools necessary for programming it.

When conducting a predictive model attack, three main stages are involved, as depicted in Fig. 5.2. The first stage entails capturing the inputs and outputs of the eFPGA to create a predictive model while also ensuring that the altered design's latency is measured for timing issues. The second stage involves finding a suitable predictive model that can be implemented on the eFPGA hardware, including model fitting, refinement, and automated multi-layer perceptron exploration. Finally, the third stage results in a synthesizable C description of the predictive model that fulfills all the given constraints while accurately approximating the behavior of the original design.

The final phase of generating a predictive model for HLS involves obtaining the smallest possible implementation of the predictive model with a latency equivalent to that of the exact version extracted in the initial phase. This is achieved by exploring various combinations of synthesis options for the optimized synthesizable predictive model and generating the most compact implementation with a latency of eFPGA (L_{efpga}). The authors discuss the use of synthesis directives (pragmas) for synthesizing arrays, loops, and functions and how different combinations of these directives result in a unique microarchitecture with specific trade-offs between area and latency. The outcome of this phase includes the pragma combination that yields the smallest predictive model implementation (pragmaopt) and the newly optimized predictive model with the exact latency as the obfuscated circuit (Copt). The final phase involves generating an eFPGA bitstream to configure an eFPGA with the predictive model.

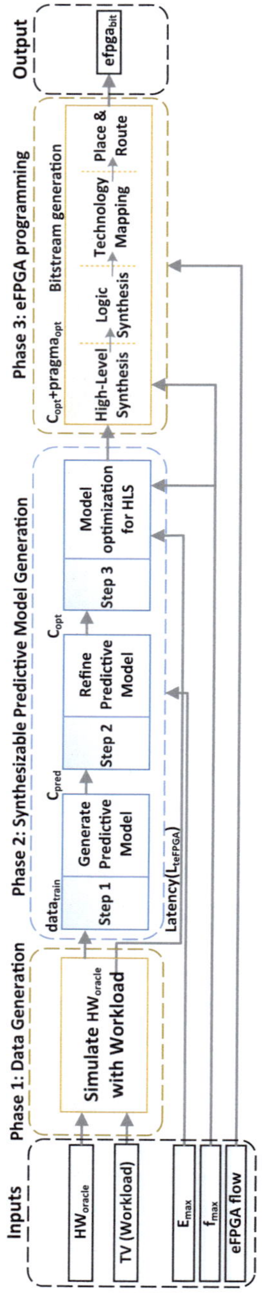

Fig. 5.2 The three primary phases of predictive model attack, adapted from [17]

However, the attack has a significant limitation as it applies only to approximate computing [17] since there is no guarantee that the reconstructed model matches the original design.

5.2.2 Break and Unroll Attack

Another attack with the name "Break and Unroll attack" has been presented to recover the bitstream of eFPGA redaction schemes. This attack is designed to be effective even when dealing with complex cycles and a large number of keys.

Researchers studying eFPGA redaction schemes find it challenging to prove their security against oracle-guided attacks. In the proposed threat model of [20–24], it is assumed that all ICs are sequential circuits, which is a reasonable assumption. It is also assumed that the attacker can access the scan-chain, which is always present, and the obfuscated netlist. This gives the adversary fine controllability/observability points within a design. As highlighted in Algorithm 2, the Break and Unroll attack encompasses two phases: cycle breaking and unrolling.

In the breaking phase of the process, all cycles in the circuit are disrupted, introducing a non-cyclic condition as a new constraint. Firstly, on line 2, all feedback signals of the circuit are identified and stored in a set called W. Then, each cycle is broken individually, and all the broken feedback signals are gathered into a CNF on line 4. This CNF is then added as a new constraint to the obfuscated circuit on line 5. Following this, a new version of the obfuscated circuit is created, with the expectation that it will not contain any structural cycles, as stated on line 6. Once the constraint is added, the SAT solver is executed on this new circuit version on line 8. However, there is a possibility that if a cycle is missed, the SAT solver could get stuck in an infinite loop. To prevent this, the *LoopDetected* function is introduced on lines 10–12, which determines whether the first part of the Break & Unroll algorithm results in an infinite loop or not. This function compares the new DIP with all members of a set that keeps all DIPs from prior iterations on lines 13-15. If the new DIP does not belong to this set, it indicates that the breaking phase operated correctly, and the correct key will be revealed in the subsequent step on lines 16–17. Conversely, if the newly generated DIP is found in the set, it signifies that the SAT solver will go into an infinite loop. Consequently, in the second phase, this issue needs to be addressed.

Algorithm 2: Break & Unroll attack algorithm [6]

Input : Obfuscated circuit $g(x, k)$ and original function $f(x)$
Output: Key vector k^* such that $g(x, k^*) \equiv f(x)$

while *(True)* **do**
  $W \leftarrow$ SearchFeedbackSignals$(g(x, k))$
  // $W \leftarrow \{w_0, w_1, \ldots, w_m\}$
  for $w_i \in W$ **do**
    $F(w_i, w_i') \leftarrow$ BreakFeedback(w_i)
    $NCCNF(k) \leftarrow \bigwedge_{i=0}^{m} F(w_i, w_i')$
    // $NCCNF(k) \leftarrow$ BreakFeedback$(w_0) \wedge \cdots \wedge$ BreakFeedback(w_m)
    $g(x, k) \leftarrow g(x, k) \wedge NCCNF(k)$
  $DIPset \leftarrow \emptyset$
  while $\hat{x} \leftarrow SAT(g(x, k_1) \neq g(x, k_2))$ **do**
    if *LoopDetected*$(\hat{x}, DIPset)$ **then**
      $w \leftarrow$ SelectFeedbackSignal(W)
      $g(x, k) \leftarrow$ NewCircuit$(w, g(x, k))$
      break
    $g(x, k_1) \leftarrow g(x, k_1) \wedge (g(\hat{x}, k_1) = f(\hat{x}))$
    $g(x, k_2) \leftarrow g(x, k_2) \wedge (g(\hat{x}, k_2) = f(\hat{x}))$
    Add$(\hat{x}, DIPset)$
  if $\neg SAT(g(x, k_1) \neq g(x, k_2))$ **then**
    return $k^* \leftarrow$ SAT$(g(x, k_1))$

The second phase, known as unrolling, aims to counteract the impact of challenging cycles in case the initial stage is unable to reveal the correct key. By addressing the limitations of the breaking phase, the unrolling process involves sequentially unrolling one cycle at a time. This includes selecting a single feedback, duplicating each gate in the circuit, and generating a new version of the obfuscated circuit. The set W, which signifies the current state of the attack algorithm, needs to be updated after each cycle unrolling. This study seeks to expose the vulnerabilities of eFPGA redaction schemes in order to emphasize the necessity of proactive measures to enhance their security. Broadly speaking, this attack capitalizes on scan-chain access, oracle availability, and SAT constructs to build the bitstream.

In practice, the attack struggles with circuits of large sizes. The results of this attack show that it is unsuccessful, as it does not work on a small circuit with only 1134 key bits. While SAT attacks are feasible, they struggle to handle circuits with large key bits, such as 100K.

5.2.3 FuncTeller Attack

In the recent "FuncTeller" attack on eFPGA redaction, researchers were able to extract the hardware IP by accessing a programmed eFPGA in a black-box manner [18]. The attack took advantage of the impact of modern EDA tools on real-world hardware circuits and used this insight to carry out the attack. The threat model outlined in this context is in line with past research on eFPGA redaction, others attacks on eFPGA, and the cybersecurity awareness worldwide competition (CSAW) [25]. The attackers in this scenario involve a collaboration between an untrusted foundry/testing facility and an untrusted end-user. The threat actors at the foundry/testing facility have the ability to access a netlist with the unprogrammed eFPGA and can isolate the eFPGA through analysis of the IC netlist or identification of the scan-chain connected to the eFPGA using reverse engineering [26]. Note that the purchased functional chip is ASIC integrated with the configured (loaded with a certain bitstream) eFPGA. This threat model is presented in detail in [18] and summarized as follows:

- One way for an attacker to isolate the eFPGA from the rest of the design is to access the dedicated eFPGA scan-chains, a feature commonly supported by eFPGA vendors [17]. Another possible method is to perform a probing attack, where the attacker locates and accesses the ASIC's internal signals to control or observe the eFPGA's inputs and outputs [17, 19].
- The extraction of hardware IP through side-channel attacks or bitstream extraction is possible [27–30].
- Once the eFPGA is isolated, an attacker can bypass scan-chain protections to enable scan-chain access. This enables the attacker to query the hardware IP design through I/O pins accessed via scan-chains. More information on the methods used to unlock or enable scan-chain access, including attacks on scan-chain protections, is presented in [31–34].

The objective of the attack is to retrieve an IP implemented on an eFPGA with only I/O access. In practical circuits, logic synthesis optimizes PPA by reducing the number of prime implicants (PIs) and literals in the circuit's prime implicants table (PIT). Consequently, ON-set minterms are grouped into multiple PIs. This consistent behavior in hardware designs has two implications. Firstly, a single PI covers multiple ON-set minterms that share the same literals in the PI representation. The authors utilize this property to predict each PI by expanding from a discovered ON-set minterm (seed). Each value is replaced by a "don't care" and heuristically verified in this process. Secondly, the hamming distance between any two PIs in a PIT is usually much smaller than the input size. This property reduces the search space when updating the predicted PIT with the new PI. Generating a new PI requires the discovery of the next ON-set minterm. Therefore, the search space for the next ON-set minterm is limited to being close to the current PIs. Utilizing the implications of circuit cones, FuncTeller employs a mechanism depicted in Fig. 5.3 to implement Boolean functions. The circuit cone has n inputs $(a_1, a_2, \ldots, a_n)$ and one output, and the example

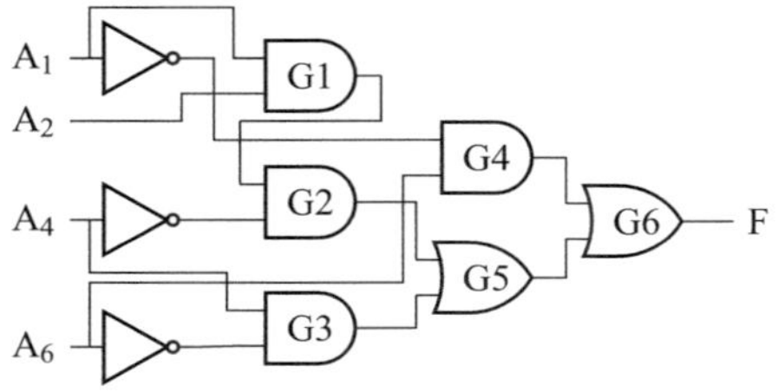

(a) The following circuit example consists of six inputs, with A_3 and A_5 left unconnected (floating).

A_1	A_2	A_3	A_4	A_5	A_6	F
1	1	–	0	–	–	1
–	–	–	1	–	0	1
0	–	–	–	–	1	1

(b) PIT Table.

Fig. 5.3 An example of a circuit and the PIT

circuit in Fig. 5.3a implements the Boolean function f with n = 6. The boolean function, $f(a_1, a_2, \ldots, a_6) = a_1a_2\overline{a_4} + a_4\overline{a_6} + \overline{a_1}a_6$, is non-canonical in its PIT representation, as displayed in Fig. 5.3b.

Algorithm 3 describes the recovery of the entire circuit. The algorithm aims to predict the entire functionality of a circuit based on an input oracle and various parameters. It takes the oracle O, which represents the circuit to be predicted, and several parameters: distance parameter d_0, linear parameter p, convergence parameter p_{conv}, and a time limit T. The algorithm's goal is to construct the predicted circuit C_{pred} that represents the entire functionality of the given circuit O. To achieve this, it utilizes several helper functions. On line 1, the main function, *predict_circuit*, begins by determining the number of outputs in the circuit using the function *count_number_of_outputs*, and stores this value in the variable $output\_size$. It initializes an empty set $Cones_{pred}$ to hold the predicted cones. On line 2, the algorithm iterates over each circuit output, represented by variable w, from 1 up to $output\_size$. On line 5, the function *predict_cone* is called to predict the cone corresponding to each output. A cone represents a part of the circuit associated with a particular output. Within *predict_cone*, the algorithm uses the distance parameter d_0, the linear parameter p, the convergence parameter p_{conv}, and the time limit T to perform the prediction. The result, denoted by PIT_{pred}, is a Predicted Information Table representing the cone's functionality. The PIT is then converted to a netlist representation using the function *convert_PIT_to_netlist* on line 6. The netlist describes the cone's structure and behavior, which is stored in the variable $Cones^{w}_{pred}$. The algorithm combines the predicted cones in the set $Cones_{pred}$ on line 7 to update the overall predicted circuit. To construct the entire predicted circuit C_{pred}, the function *merge_cones_to_circuit* takes this set as input on line 8. Finally, on line 9, the algorithm outputs the predicted circuit C_{pred}.

Algorithm 3: Predicting entire circuit's functionality [18]

Input : Oracle O, distance parameter d_0, linear parameter p, convergence parameter p_{conv}, time limit T

Output: Entire predicted circuit C_{pred}

```
Function predict_circuit(O, d0, p, p_conv, T):
    output_size := count_number_of_outputs(O);
    Cones_pred := Ø;
    for w ← 1 to output_size do
        PIT_pred := predict_cone(O, w, d0, p, p_conv, T);
        Cones^w_pred := convert_PIT_to_netlist(PIT_pred);
        Cones_pred := Cones_pred ∪ Cones^w_pred;
    C_pred := merge_cones_to_circuit(Cones_pred);
    return C_pred;
```

This attack leverages the aforementioned threat model to predict the bitstream of eFPGA. However, for some small circuits like $c1355$ and $c1908$, the attack could not find the key. Additionally, the attack assumes that the eFPGA IP can be easily separated from the obfuscated design. While these attacks aim to break eFPGA redaction, none can completely recover the bitstream. Instead, they predict the bitstream, and there is a probability that the predicted bitstream may be correct to some degree. Yet, there is no obvious way for an adversary to know how far from the real correct solution he/she is.

5.3 Security Analysis of ReBO

The security analysis is an essential component of obfuscation. This section will cover the security analysis conducted using three major types of ReBO techniques. Despite their ability to resist attacks, the high costs associated with ReBO techniques may render them impractical. Overall, most of the research on ReBO claims that the proposed techniques are resistant to SAT. Many authors have utilized the SAT attack to evaluate their proposed techniques (e.g., [35–38]). The summary of attacks executed to evaluate the security of various techniques is presented in Table 5.1.

As indicated in Table 5.1, some authors solely employed security analysis (SA) to assess the effectiveness of their defense techniques. If authors effectively attempted to attack their own defenses, we marked them as applied attacks (AA). Here, we make a clear argument that both SA and AA have merits, but SA can be easily misinterpreted to suggest the techniques are more secure than they actually are. For instance, a classical SA discussion is the enumeration of the adversarial search space. However, even large search spaces can be broken, and this can only be shown through AA.

Table 5.1 Security comparisons of FPGA-inspired obfuscation techniques

Technique		Attack resiliency			
		SA versus AA*	SAT	PSCA	Others
LUT-based	PICPRLB [39]	–	*No*	–	–
	eREL [40]	AA	*No*	–	Hardware-based code and injection attack
	LUT-Lock [41]	AA	*Yes*	–	–
	CAD-hASIC [42]	AA	*Yes*	–	AppSAT, removal attack, composition attack, structural attack, and SCOPE
eFPGA redaction	SILICON-LUT [43]	AA	*Yes*	–	Removal attack
	ALICE [44]	–	–	–	–
	eFPGA REDAC [22]	AA	–	–	Icy-SAT
	FINE-GRAIN eFPGA [21]	AA	*Yes*	–	–
	eFPGA-PARTIAL DESIGN [35]	AA	*Yes*	–	Brute force attack
	eFPGA-IP REDAC [45]	AA	*Yes*	–	Icy-SAT, CycSAT, and Be-SAT
	CGRRA [46]	SA	*Yes*	–	Removal attack
	SAT RES-RECOFIG [47]	AA	*Yes*	–	Removal attack
	TRAP [38]	SA	*Yes*	–	Brute force attack
Emerging technologies	MTJ-STT-LUT [48]	AA	*Yes*	–	Removal attack, scan-based attack, ATPG-based attack, approximate attack, and SMT-based attack
	SPS-SHE [49]	AA	*Yes*	–	Double DIP
	DYNAMIC-RILB [50]	AA	*Yes*	*Yes*	AppSAT, removal attack, scan and shift-based attack
	LOCK&ROLL [51]	AA	*Yes*	*Yes*	Removal attack, scan and shift-based attack
	Hybrid STT-LUT[52]	SA	*No*	–	Brute force attack, machine learning-based attack
	SOT-LUT [53]	SA	–	–	Brute force attack, side-channel based attack, testing-based attack, circuit partition-based attack
	LUT-OBF [54]	AA	*Yes*	*Yes*	–
	SECURITY LUT-OBF [55]	AA	*Yes*	–	AppSAT, removal attack, scan, and shift-based attack
	STT MAGNETIC-LUT [56]	–	*No*	–	–

*SA and AA are abbreviations for security analysis and applied attack

The effectiveness of the techniques was mainly assessed against SAT attacks, as detailed in the fourth column of Table 5.1. However, it is important to note that most of the techniques have not been tested against power side-channel attacks (PSCAs), especially in the case of CMOS implementations. Regarding LUT-based obfuscation, the authors of [40, 42] have conducted a variety of attacks. eFPGA redaction has been subjected to the majority of SAT and removal attacks for security assessment. The category of obfuscation with emerging technologies is vulnerable to various attacks, including brute-force, testing-based, and side-channel attacks. The authors of [53] and [38] deliberated on the robustness of their methods against testing-based attacks, circuit partition-based attacks, brute force attacks, SAT attacks, and PSCAs. The "Others" column in the table reflects the resilience of a particular method against other types of attacks. It has been repeatedly observed that LL mechanisms, which were initially considered secure—and in some cases, proven to be so—have succumbed to seemingly "simple" attacks that were not initially considered.

One of the main obstacles involves establishing a security measurement for ReBO techniques. Currently, there is no widely adopted common security metric. In order to impartially evaluate the security of ReBO techniques, we utilized a standard criterion: the ratio of gates to the length of the bitstream. In this context, "gates" refers to logic gates similar to those used in logic synthesis, indicating that a gate represents a standard cell. For example, if an author determined that a circuit with additional 1000 key gates was sufficiently obfuscated with a 128-bit bitstream, the metric for this circuit would be 7.81. Typically, the level of obfuscation or the concealed portion tends to increase in relation to the length of the bitstream. Developers of defensive techniques aim to maintain a lower value of the metric in order to ensure the higher security of a circuit.

In many cases, authors discuss the area of obfuscated design, and based on that, we can estimate the number of gates from the area using information about the technology used for implementation. In the comparison shown in Fig. 5.4, the gates count to bitstream size ratio of benchmark circuits serves as a security metric. We calculated the minimum, maximum, and average values of the ratio in most of the techniques[1] considering all benchmarks and bitstream sizes used originally by the developers for evaluating their techniques. In general, a high ratio usually indicates more expensive security [38, 41, 49]. However, there are exceptions to this rule [41]. Moreover, optimizing a design to achieve desirable PPA overheads along with security would make the relation between security and bitstream size more complex [45]. In some cases, calculating the ratio was simple because the authors provided the gate count and the size of the bitstream [24, 51, 55]. However, for other techniques, the ratio could not be calculated due to missing values [39, 41], or estimations were needed for the calculation.

In the eFPGA context, there are two conflicting approaches to security. Some techniques suggest that obfuscating around 10% to 30% of a circuit's gate count is sufficient to thwart SAT attacks, while others, such as those discussed in [24], recommend obfuscation rates of over 86% to protect more than the functionality of the circuit, but also its intent. As shown

[1] Not all surveyed techniques provide enough information for this comparison, unfortunately.

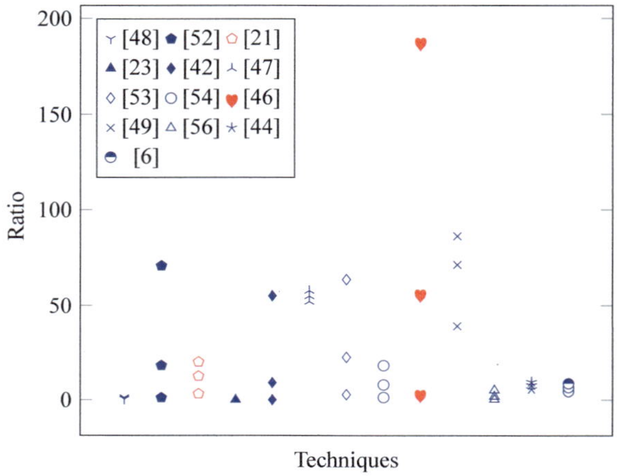

Fig. 5.4 Ratio of number of gates to the bitstream length for several techniques. The red color indicates that corresponding data was estimated, whereas the blue color indicates that the data was taken directly from related references

in Fig. 5.4, the first solution results in a significantly lower obfuscation ratio compared to the second, but it is important to note that the objectives and threat models of these techniques differ.

In this chapter, we discussed the various classes of attacks that are prevented by LL. We also talked about the emerging attacks targeting ReBO. Another aspect of ReBO we covered was the evolution of security against numerous attacks. We took this into consideration and discussed LUT-based obfuscation, eFPGA redaction, and obfuscation with emerging technologies. In the next three chapters, we will describe the obfuscation methodologies within these categories.

References

1. P. Subramanyan, S. Ray, and S. Malik, "Evaluating the security of logic encryption algorithms," in *2015 IEEE International Symposium on Hardware Oriented Security and Trust (HOST)*, pp. 137–143, 2015.
2. J. Rajendran, Y. Pino, O. Sinanoglu, and R. Karri, "Security analysis of logic obfuscation," in *Design Automation Conference*, pp. 83–89, 2012.
3. H. Zhou, R. Jiang, and S. Kong, "CycSAT: SAT-based attack on cyclic logic encryptions," in *2017 IEEE/ACM International Conference on Computer-Aided Design (ICCAD)*, pp. 49–56, 2017.
4. K. Shamsi, M. Li, T. Meade, Z. Zhao, D. Z. Pan, and Y. Jin, "AppSAT: Approximately deobfuscating integrated circuits," in *2017 IEEE International Symposium on Hardware Oriented Security and Trust (HOST)*, pp. 95–100, 2017.
5. K. Z. Azar, H. M. Kamali, H. Homayoun, and A. Sasan, "SMT attack: Next generation attack on obfuscated circuits with capabilities and performance beyond the SAT attacks," *IACR Transactions on Cryptographic Hardware and Embedded Systems*, pp. 97–122, 2019.

6. A. Rezaei, R. Afsharmazayejani, and J. Maynard, "Evaluating the security of eFPGA-based redaction algorithms," ICCAD '22, (New York, NY, USA), Association for Computing Machinery, 2022.
7. M. Yasin, B. Mazumdar, O. Sinanoglu, and J. Rajendran, "Removal attacks on logic locking and camouflaging techniques," *IEEE Transactions on Emerging Topics in Computing*, vol. 8, no. 2, pp. 517–532, 2020.
8. Y. Shen and H. Zhou, "Double dip: Re-evaluating security of logic encryption algorithms," in *Proceedings of the Great Lakes Symposium on VLSI 2017*, GLSVLSI '17, p. 179–184, 2017.
9. Z. U. Abideen, S. Gokulanathan, M. J. Aljafar, and S. Pagliarini, "An overview of FPGA-inspired obfuscation techniques," *Association for Computing Machinery*, vol. 56, no. 12, December 2024.
10. Y.-W. Lee and N. A. Touba, "Improving logic obfuscation via logic cone analysis," in *2015 16th Latin-American Test Symposium (LATS)*, pp. 1–6, 2015.
11. M. Yasin, B. Mazumdar, O. Sinanoglu, and J. Rajendran, "Security analysis of anti-sat," in *2017 22nd Asia and South Pacific Design Automation Conference (ASP-DAC)*, pp. 342–347, 2017.
12. P. Chakraborty, J. Cruz, A. Alaql, and S. Bhunia, "SAIL: Analyzing structural artifacts of logic locking using machine learning," *IEEE Transactions on Information Forensics and Security*, pp. 1–1, 2021.
13. A. Alaql, D. Forte, and S. Bhunia, "Sweep to the secret: A constant propagation attack on logic locking," in *2019 Asian Hardware Oriented Security and Trust Symposium (AsianHOST)*, pp. 1–6, 2019.
14. A. Alaql, M. M. Rahman, and S. Bhunia, "SCOPE: Synthesis-based constant propagation attack on logic locking," *IEEE Transactions on Very Large Scale Integration (VLSI) Systems*, vol. 29, no. 8, pp. 1529–1542, 2021.
15. L. Alrahis, S. Patnaik, F. Khalid, M. A. Hanif, H. Saleh, M. Shafique, and O. Sinanoglu, "Gnnunlock: Graph neural networks-based oracle-less unlocking scheme for provably secure logic locking," in *2021 Design, Automation & Test in Europe Conference & Exhibition (DATE)*, pp. 780–785, 2021.
16. Y. Xie and A. Srivastava, "Anti-sat: Mitigating sat attack on logic locking," *IEEE Transactions on Computer-Aided Design of Integrated Circuits and Systems*, vol. 38, no. 2, pp. 199–207, 2019.
17. P. Chowdhury, C. Sathe, and B. Carrion Schaefer, "Predictive model attack for embedded FPGA logic locking," in *Proceedings of the ACM/IEEE International Symposium on Low Power Electronics and Design*, pp. 1–6, 2022.
18. Z. Han, M. Shayan, A. Dixit, M. Shihab, Y. Makris, and J. J. Rajendran, "FuncTeller: How well does eFPGA hide functionality?," in *32nd USENIX Security Symposium (USENIX Security 23)*, pp. 5809–5826, 2023.
19. H. Wang, D. Forte, M. M. Tehranipoor, and Q. Shi, "Probing attacks on integrated circuits: Challenges and research opportunities," *IEEE Design & Test*, vol. 34, no. 5, pp. 63–71, 2017.
20. L. Collini, B. Tan, C. Pilato, and R. Karri, "Reconfigurable logic for hardware IP protection: Opportunities and challenges," in *Proceedings of the 41st IEEE/ACM International Conference on Computer-Aided Design*, pp. 1–7, 2022.
21. P. Mohan, O. Atli, J. Sweeney, O. Kibar, L. Pileggi, and K. Mai, "Hardware redaction via designer-directed fine-grained eFPGA insertion," in *2021 Design, Automation & Test in Europe Conference & Exhibition (DATE)*, pp. 1186–1191, IEEE, 2021.
22. J. Bhandari, A. K. Thalakkattu Moosa, B. Tan, C. Pilato, G. Gore, X. Tang, S. Temple, P.-E. Gaillardon, and R. Karri, "Exploring eFPGA-based redaction for IP protection," in *2021 IEEE/ACM International Conference On Computer Aided Design (ICCAD)*, pp. 1–9, 2021.
23. J. Bhandari, A. K. T. Moosa, B. Tan, C. Pilato, G. Gore, X. Tang, S. Temple, P.-E. Gaillardo, and R. Karri, "Not all fabrics are created equal: Exploring eFPGA parameters for IP redaction," *arXiv preprint*, arXiv:2111.04222, 2021.

24. Z. U. Abideen, T. D. Perez, and S. Pagliarini, "From FPGAs to obfuscated eASICs: Design and security trade-offs," in *2021 Asian Hardware Oriented Security and Trust Symposium (AsianHOST)*, pp. 1–4, 2021.
25. K. Ramesh, B. Tan, L. Collini, and T. M. A. Khader, "CSAW'21 logic locking," last accessed on Jul 27, 2023. Available at: https://www.csaw.io/logic-locking.
26. R. Torrance and D. James, "The state-of-the-art in IC reverse engineering," in *Cryptographic Hardware and Embedded Systems - CHES 2009* (C. Clavier and K. Gaj, eds.), (Berlin, Heidelberg), pp. 363–381, Springer Berlin Heidelberg, 2009.
27. O. Glamočanin, D. G. Mahmoud, F. Regazzoni, and M. Stojilović, "Shared FPGAs and the holy grail: Protections against side-channel and fault attacks," in *2021 Design, Automation & Test in Europe Conference & Exhibition (DATE)*, pp. 1645–1650, 2021.
28. J. Szefer, "Survey of microarchitectural side and covert channels, attacks, and defenses," *Journal of Hardware and Systems Security*, vol. 3, pp. 219–234, Sep 2019.
29. K. Tiri and I. Verbauwhede, "A logic level design methodology for a secure DPA resistant ASIC or FPGA implementation," in *Proceedings Design, Automation and Test in Europe Conference and Exhibition*, vol. 1, pp. 246–251 Vol.1, 2004.
30. B. Yang, K. Wu, and R. Karri, "Scan based side channel attack on dedicated hardware implementations of data encryption standard," in *2004 International Conferce on Test*, pp. 339–344, 2004.
31. S. S. Ali, S. M. Saeed, O. Sinanoglu, and R. Karri, "Novel test-mode-only scan attack and countermeasure for compression-based scan architectures," *IEEE Transactions on Computer-Aided Design of Integrated Circuits and Systems*, vol. 34, no. 5, pp. 808–821, 2015.
32. L. Alrahis, M. Yasin, N. Limaye, H. Saleh, B. Mohammad, M. Al-Qutayri, and O. Sinanoglu, "ScanSAT: Unlocking static and dynamic scan obfuscation," *IEEE Transactions on Emerging Topics in Computing*, vol. 9, no. 4, pp. 1867–1882, 2021.
33. J. DaRolt, G. Di Natale, M.-L. Flottes, and B. Rouzeyre, "Scan attacks and countermeasures in presence of scan response compactors," in *2011 Sixteenth IEEE European Test Symposium*, pp. 19–24, 2011.
34. N. Limaye and O. Sinanoglu, "DynUnlock: Unlocking scan chains obfuscated using dynamic keys," in *2021 Design, Automation & Test in Europe Conference & Exhibition (DATE)*, p. 270–273, 2020.
35. B. Hu, T. Jingxiang, S. Mustafa, R. R. Gaurav, S. William, M. Yiorgos, C. S. Benjamin, and S. Carl, "Functional obfuscation of hardware accelerators through selective partial design extraction onto an embedded FPGA," in *Proceedings of the 2019 Great Lakes Symposium on VLSI*, p. 171–176, 2019.
36. P. Subramanyan, S. Ray, and S. Malik, "Evaluating the security of logic encryption algorithms," in *2015 IEEE International Symposium on Hardware Oriented Security and Trust (HOST)*, pp. 137–143, 2015.
37. J. Chen, M. Zaman, Y. Makris, R. D. S. Blanton, S. Mitra, and B. C. Schafer, "DECOY: Deflection-Driven HLS-Based Computation Partitioning for Obfuscating Intellectual PropertY," in *Proceedings of the 57th ACM/EDAC/IEEE Design Automation Conference*, DAC '20, IEEE Press, 2020.
38. M. M. Shihab, J. Tian, G. R. Reddy, B. Hu, W. Swartz, B. Carrion Schaefer, C. Sechen, and Y. Makris, "Design obfuscation through selective post-fabrication transistor-level programming," in *2019 Design, Automation & Test in Europe Conference & Exhibition (DATE)*, pp. 528–533, 2019.
39. A. Baumgarten, A. Tyagi, and J. Zambreno, "Preventing IC piracy using reconfigurable logic barriers," *IEEE Design Test of Computers*, vol. 27, no. 1, pp. 66–75, 2010.

40. B. Liu and B. Wang, "Embedded reconfigurable logic for ASIC design obfuscation against supply chain attacks," in *2014 Design, Automation Test in Europe Conference Exhibition (DATE)*, pp. 1–6, 2014.
41. H. Mardani Kamali, K. Zamiri Azar, K. Gaj, H. Homayoun, and A. Sasan, "LUT-lock: A novel LUT-based logic obfuscation for FPGA-bitstream and ASIC-hardware protection," in *2018 IEEE Computer Society Annual Symposium on VLSI (ISVLSI)*, pp. 405–410, 2018.
42. Z. U. Abideen, T. D. Perez, M. Martins, and S. Pagliarini, "A security-aware and lut-based cad flow for the physical synthesis of hasics," *IEEE Transactions on Computer-Aided Design of Integrated Circuits and Systems*, vol. 42, no. 10, pp. 3157–3170, 2023.
43. G. Kolhe, T. Sheaves, K. I. Gubbi, T. Kadale, S. Rafatirad, S. M. PD, A. Sasan, H. Mahmoodi, and H. Homayoun, "Silicon validation of LUT-based logic-locked IP cores," in *Proceedings of the 59th ACM/IEEE Design Automation Conference*, pp. 1189–1194, 2022.
44. C. M. Tomajoli, L. Collini, J. Bhandari, A. K. T. Moosa, B. Tan, X. Tang, P.-E. Gaillardon, R. Karri, and C. Pilato, "ALICE: An automatic design flow for eFPGA redaction," in *Proceedings of the 59th ACM/IEEE Design Automation Conference*, p. 781–786, 2022.
45. J. Bhandari, A. K. T. Moosa, B. Tan, C. Pilato, G. Gore, X. Tang, S. Temple, P.-E. Gaillardon, and R. Karri, "Not all fabrics are created equal: Exploring efpga parameters for ip redaction," *IEEE Trans. Very Large Scale Integr. Syst.*, vol. 31, p. 1459–1471, oct 2023.
46. J. Chen and B. C. Schafer, "Area efficient functional locking through coarse grained runtime reconfigurable architectures," in *Proceedings of the 26th Asia and South Pacific Design Automation Conference*, pp. 542–547, 2021.
47. S. D. Chowdhury, G. Zhang, Y. Hu, and P. Nuzzo, "Enhancing SAT-attack resiliency and cost-effectiveness of reconfigurable-logic-based circuit obfuscation," in *2021 IEEE International Symposium on Circuits and Systems (ISCAS)*, pp. 1–5, IEEE, 2021.
48. G. Kolhe, T. D. Sheaves, S. M. P. D., H. Mahmoodi, S. Rafatirad, A. Sasan, and H. Homayoun, "Breaking the design and security trade-off of look-up table-based obfuscation," *ACM Trans. Des. Autom. Electron. Syst.*, 2022.
49. S. Patnaik, N. Rangarajan, J. Knechtel, O. Sinanoglu, and S. Rakheja, "Advancing hardware security using polymorphic and stochastic spin-hall effect devices," in *2018 Design, Automation & Test in Europe Conference & Exhibition (DATE)*, pp. 97–102, IEEE, 2018.
50. G. Kolhe, S. Salehi, T. D. Sheaves, H. Homayoun, S. Rafatirad, M. P. Sai, and A. Sasan, "Securing hardware via dynamic obfuscation utilizing reconfigurable interconnect and logic blocks," in *2021 58th ACM/IEEE Design Automation Conference (DAC)*, pp. 229–234, IEEE, 2021.
51. G. Kolhe, T. Sheaves, K. I. Gubbi, S. Salehi, S. Rafatirad, S. M. PD, A. Sasan, and H. Homayoun, "Lock&roll: deep-learning power side-channel attack mitigation using emerging reconfigurable devices and logic locking," in *Proceedings of the 59th ACM/IEEE Design Automation Conference*, pp. 85–90, 2022.
52. T. Winograd, H. Salmani, H. Mahmoodi, K. Gaj, and H. Homayoun, "Hybrid STT-CMOS designs for reverse-engineering prevention," in *Proceedings of the 53rd Annual Design Automation Conference*, pp. 1–6, 2016.
53. J. Yang, X. Wang, Q. Zhou, Z. Wang, H. Li, Y. Chen, and W. Zhao, "Exploiting spin-orbit torque devices as reconfigurable logic for circuit obfuscation," *IEEE Transactions on Computer-Aided Design of Integrated Circuits and Systems*, vol. 38, no. 1, pp. 57–69, 2018.
54. G. Kolhe, S. M. PD, S. Rafatirad, H. Mahmoodi, A. Sasan, and H. Homayoun, "On custom LUT-based obfuscation," in *Proceedings of the 2019 on Great Lakes Symposium on VLSI*, GLSVLSI '19, p. 477–482, Association for Computing Machinery, 2019.
55. G. Kolhe, H. M. Kamali, M. Naicker, T. D. Sheaves, H. Mahmoodi, P. D. Sai Manoj, H. Homayoun, S. Rafatirad, and A. Sasan, "Security and complexity analysis of LUT-based obfuscation:

From blueprint to reality," in *2019 IEEE/ACM International Conference on Computer-Aided Design (ICCAD)*, pp. 1–8, 2019.
56. A. Attaran, T. D. Sheaves, P. K. Mugula, and H. Mahmoodi, "Static design of spin transfer torques magnetic look up tables for ASIC designs," in *Proceedings of the 2018 on Great Lakes Symposium on VLSI*, pp. 507–510, 2018.
57. N. Rangarajan, S. Patnaik, J. Knechtel, R. Karri, O. Sinanoglu, and S. Rakheja, "Opening the doors to dynamic camouflaging: Harnessing the power of polymorphic devices," *IEEE Transactions on Emerging Topics in Computing*, 2020.

6 LUT-Based Obfuscation

6.1 Obfuscation Using Reconfigurable Logic Barriers

After the emergence of LL in [1], the first idea of using reconfigurable logic came in 2010 [2, 3]. This scheme decomposes IP functionality $F(x)$ into F_{fixed} and F_{reconfig}. The complete process of generating the reconfigurable barriers is illustrated in Fig. 6.1. The majority of the design, F_{fixed}, is given to the foundry to fabricate, but F_{reconfig} remains with the IP creator. Rather than fabricating F_{reconfig} as the original logic, the idea is to fabricate F_{reconfig} as reconfigurable logic.[1] Using a secure key distribution framework, the withheld F_{reconfig} partition can be programmed into the reconfigurable locations during the activation stage. An adversarial foundry can make extra ICs; however, without knowledge of the withheld configuration, the ICs would not function correctly. The foundry could also guess the configuration, but to make an educated decision regarding the reconfigurable locations, an adversary would have to invest more time and resources than are practical to discover the correct configuration.

Due to the key's significance, a distribution framework is needed to securely unlock each IC. The transmission of F_{reconfig} from the IC designer to the IC is secured, and ICs can be activated individually. Controllability and observability are metrics used for the selection of reconfigurable barriers. Controllability measures how easy it is to control a node's inputs, while observability measures how easy it is to see specific values at internal nodes on observable outputs. These metrics are often expressed in terms of don't-care minterm sets CDC_{in} (controllability don't-care) and ODC_{out} (observability don't-care).

The locations to insert reconfigurable barriers representing F_{reconfig} are chosen to minimize the possibility of the attacker bypassing the locks or guessing the correct configuration. The best cut for logic barriers is selected, considering the ODC set and cut height. ODC

[1] Also known as LUT or reconfigurable barrier.

Z. U. Abideen and S. Pagliarini, *Reconfigurable Obfuscation Techniques for the IC Supply Chain*, Synthesis Lectures on Digital Circuits & Systems,
https://doi.org/10.1007/978-3-031-77509-3_6

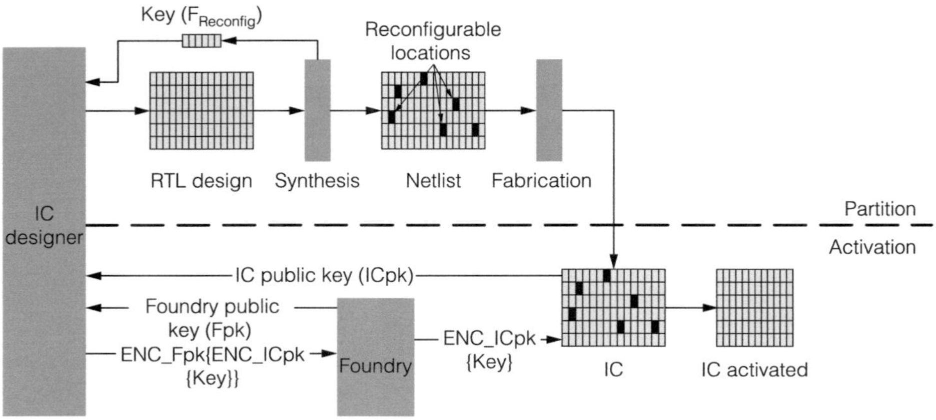

Fig. 6.1 The design is divided before fabricating to generate a key that is securely distributed to the IC for activation [2]

is used because it maximizes the control by being highly observable over outputs. Formally, for the ogic barrier $N = \{V, E\}$, corresponding to a function $f(x_0, x_1, \ldots, x_{n-1}) = (y_0, y_1, \ldots, y_{N-1})$ with the desired threshold of reconfigurable logic, $0 \leq \text{TH}_{\text{reconfig}} < 1$, a subset $V_{\text{reconfig}} \subseteq V$ is chosen such that $V_{\text{reconfig}}/V \leq TH_{\text{reconfig}}$ and the observability of subunits G_i in f is maximized by computing $\sum_{G_i \in V_{\text{reconfig}}} \sum_{j=0}^{N-1} \min(x_0, x_1, \ldots, x_{N-1}) \frac{\partial y_i}{\partial G_i}$.

The height is also used as a metric when selecting cuts for the logic barriers. Cuts near the center are prioritized, while cuts near primary outputs are highly observable but less controllable, and cuts near primary inputs have the opposite characteristics. During the cut selection of gates, the parameter α was introduced to balance the *ODC* minterm score and height value. An α value of 0 considers only the *ODC* score, while a value of 100 relies completely on height.

6.1.1 Security Analysis

Concerning the security analysis, the considered circuits are affected by a single-cut heuristic. The impact of a single-cut heuristic on benchmarks was considered a security. The Hamming distance versus the percentage of gates in a single-cut selection was represented. A security metric of Hamming distances for a single cut that was consistently close to 50% was defined, and the percentage of gates selected was very low for most benchmarks, averaging approximately 8%. It was suggested that a single cut effectively makes the circuit act as random logic while requiring only a small percentage of reconfigurable logic.

The effects of a cut's height in selection heuristics were introduced with a parameter '*a*' to weigh the heuristic's dependence on *ODC* and height values. The range of '*a*' varied from 0 to 100, resulting in fewer than 101 unique netlists. Each unique netlist was simulated under the assumption of an intelligent attacker. The attacker measured the percentage of correct outputs for each test vector and treated the LUTs as independent. LUT configurations were constant for all LUTs except the one iterated over. The configuration leading to the greatest information gain was chosen after applying millions of test vectors and monitoring the outputs. This process continued through all LUTs, and an equal balance between the *ODC* and height created greater difficulty for an attacker as they did not gain as much information. Additionally, no additional information is gained as the effort increases, making this approach resilient against attacks.

6.2 LUT-Lock

LUT-Lock [4] is a technique that obfuscates a netlist while embedding several key features to make the obfuscation a hard problem for state-of-the-art attacks, with a focus on SAT attacks [5]. The developers have designed a defense mechanism that identifies key features to increase the difficulty of obfuscation for SAT attacks. The LUT-Lock algorithm combines three features to provide the best defense against SAT attacks.

The *first* selection method selects candidate gates based on the observation that higher output corruption reduces the resiliency of obfuscation solutions against SAT attacks. This is achieved by limiting LUT insertion to the fan-in cone of the smallest possible set of primary outputs as "fan-In cone (FIC)". To enhance obfuscation strength, LUTs are mapped to affect the minimum number of primary outputs (POs), thereby reducing output corruption. When performing LUT replacement using FIC, it is important to prioritize replacing cells closest to the selected output using breadth-first search (BFS). The delay of all timing paths through a candidate gate should be estimated to avoid timing violations, too. If the delay exceeds a predefined threshold, the replacement is discarded and the next candidate is checked. After all gates in the current fan-in cone are replaced, a new primary output is selected. The output pins selected for obfuscation must satisfy two conditions: the total positive slack (TPS) of all timing paths leading to the primary output(s) should be large, and it must have a large fan-in cone size. For large circuits, coefficients α and β are defined to prioritize these conditions. The cumulative weight to aid in selecting the best candidate output is obtained using $\alpha \cdot \text{TPS}^* + \beta \cdot \text{FIC}^*$.[2]

In the *second* selection method, the gates with higher signal probability skew (SPS) at their output are considered better candidates for obfuscation. The gate selection strategy in the proposed algorithm is referred to as "focusing on higher skew gates (HSC)" within the fan-in cone of selected outputs based on FIC. The replacement priority is given to gates with higher SPS. When a gate is selected for obfuscation, its fan-in gates will be added to the list

[2] The values *TPS* and FIC** in the circuit are normalized relative to their maximum potential values.

of gates that could be visited in the next search for gate replacement, and the gates with the highest SPS will be selected from this list. Each gate replacement candidate must pass the timing check; otherwise, it will be ignored.

To reduce output corruption and make SAT attacks more difficult, the third method can be prioritized for obfuscation for gates with the lowest fan-out, which is a *third* method, referred to as "focusing on gates with minimum fan-out (MO-HSC)". FIC employs a BFS to visit all candidate gates and then selects gates with the minimum number of fan-outs. In case of a tie, the gate with the highest SPS is chosen. When a gate is obfuscated, its fan-in gates are added to the list of candidate gates for the next selection. Just like FIC, replacements must pass a timing check to be considered. This approach minimizes primary output corruption without significantly increasing the area.

In the *fourth* method, the gates with multiple fan-outs are selected but only affect one output, which is referred to as "focusing on gates with the least impact on POs (MFO-HSC)". FLIP focuses on the number of outputs connected to each candidate gate, reducing output corruption resulting from obfuscation. Candidate gate ties are resolved using the SPS of their respective gates. When a gate is selected for obfuscation, its fan-in gates are added to the list of candidate gates for the next selection. Furthermore, any gate replacement candidate must pass the timing check, or else it will be ignored.

Next, the focus is on avoiding back-to-back insertion of LUTs, which is referred to as (NB2-MO-HSC)". In this case, the obfuscation of gates with LUTs is affected by increased key possibilities due to the freedom to exploit gate conversion based on De Morgan's laws. For instance, back-to-back obfuscation of the function $(A \vee B) \wedge (C \vee D)$ using 2-input LUTs can result in four different combinations of programmable logic based on De Morgan's laws. The number of correct keys increases as more gates are added to the logic cone, leading to an exponential increase in the number of valid keys. This is known as correct key explosion. Obfuscating more gates may reduce the obfuscation strength.

6.2.1 Security Analysis

The effectiveness of each main aspect of the LUT-Lock is showcased by using the execution time of a SAT solver. LUT-Lock is also compared to STT-MTJ based LUT [6] and reconfigurable barriers [2]. The SAT solver's execution time is increased as the replacement algorithm evolves from random replacement to FIC, to HSC, to MO-HSC, to MFO-HSC, and to NB2-MO-HSC (LUT-Lock). This progression illustrates the orthogonal improvement of added features in providing resiliency against SAT attacks. The LUT-Lock algorithm with the NB2-MO-HSC replacement policy significantly increases the SAT attack execution time as the number of obfuscated gates grows. This is illustrated in Fig. 6.2, where the execution time of an SAT solver and the number of generated keys per each inserted LUT for the benchmark *c*5315 is plotted. The LUTs are placed back-to-back. Hence, inserting each LUT increases the number of keys.

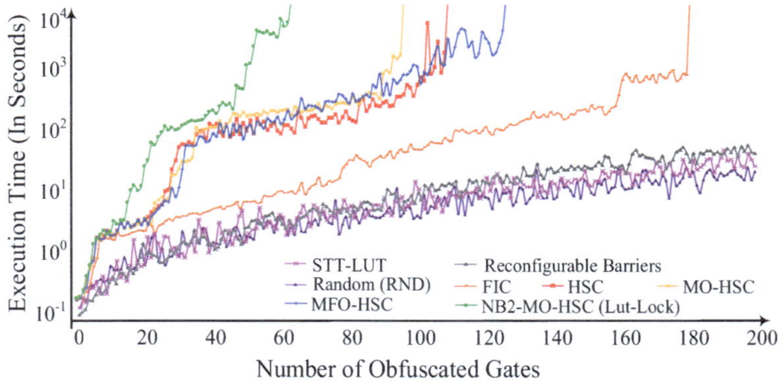

Fig. 6.2 Execution time of the SAT attack for finding a valid key when using LUT-Lock (NB2-MO-HSC) compared to its sub-algorithms (RND, FIC, HSC, MO-HSC, and MFO-HSC) and the work in [6] and [2]. The considered circuit is *c*5315 [4]

From Fig. 6.2, the SAT resiliency of prior work is close to that of random replacement, showing slow growth in SAT attack execution time with respect to the number of inserted gates. In contrast, the LUT-Lock replacement method clearly shows a much faster exponential increase in difficulty. With only 20 replaced LUTs, the LUT-Lock obfuscated netlist is as resilient as the netlist produced by the [6] and [2] replacement policies when using 10 times (200 gates) the number of gates. By increasing the number of gates, the SAT resiliency of the LUT-Lock insertion policy continues to grow exponentially.

6.3 Full-Lock

In previous discussions, we focused on obfuscation implemented with LUTs. Now, we switch to solutions involving switch boxes (SwB) and routing circuitry alongside LUTs. "Full-Lock" creates complex and time-consuming computational models to thwart the Davis–Putnam–Logemann–Loveland (DPLL) algorithm [7].

In Full-Lock, Mux gates provide a better building block for constructing SAT-hard circuits than XOR/XNOR gates. This is because Mux gates, with their four variables, can make the recursive DPLL tree one level deeper, requiring more pruning and backtracking [8]. A switching network using Muxes can be constructed to complicate the circuit further and increase the SAT hardness to the desired range. This network prevents variables from being resolved until all cascaded variables are resolved, which stops the circuit from simplifying before reaching the decision tree's leaves. N-by-M switch-boxes with back-to-back interconnections have created challenging instances that trap even the best solvers in difficult solution spaces for extended periods before finding a satisfying solution.

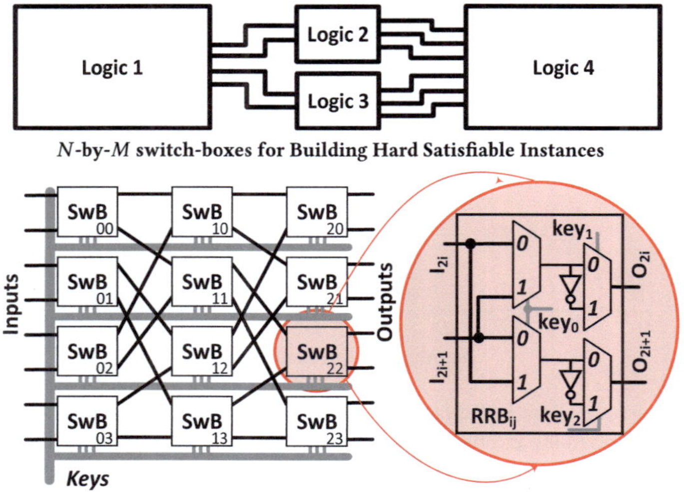

Fig. 6.3 Shuffle-based CLN with N = 8 [7]

The key-configurable logarithmic-based network (CLN) is constructed in Full-Lock to obfuscate routes. For this purpose, small and lightweight SwB are created and easily implemented using only Muxes, as shown in Fig. 6.3. These small and lightweight SwB allow a large logarithmic switching ($\log_2$N) network to be created and benefit from the hardness of switch-boxes while the power, performance, and area overhead remain reasonable.

A set of self-routing logarithmic networks, $\log_2$N networks, is provided across all switching networks, offering configurable interconnection with less overhead than conventional networks such as mesh or crossbar. A simple implementation of an 8×8 CLN using the blocking shuffle network is demonstrated in Fig. 6.3 [9]. This CLN is constructed using small SwB, where each SwB is built using Muxes. In each switch box, the outputs can be an arbitrary permutation of the inputs. Additionally, key-configurable inverters are added for each wire, allowing an output to be shuffled and negated based on the key value. The CLN has N inputs, and N is a power of 2 due to its structure. The number of SwBs in a CLN depends on the number of inputs as well as the model of $\log_2 N$ networks.

In order to significantly increase the number of possible permutations without adding extra space, a near non-blocking[3] logarithmic network proposed in [10] was utilized to create a key-configurable CLN. Almost non-blocking performance can be achieved by this network, and nearly all permutations can be handled using a $\text{LOG}_{N,\log_2(N)-2,1}$ configuration, with only $\log_2(N) - 2$ extra stages and no additional duplicates. As shown, the structure of SwB interconnections differs from the shuffle-based topology depicted in Fig. 6.3.

[3] In non-blocking logarithmic networks like $\text{LOG}_{N,M,P}$, N represents the number of inputs/outputs, M represents the number of extra stages, and P represents additional vertically cascaded copies.

When comparing an almost non-blocking CLN to a blocking CLN with the same number of inputs, the almost non-blocking CLN has roughly two times the area and power overhead, with only $\log_2(N) - 2$ extra stages. However, it offers much greater resistance to SAT attacks. For instance, a non-blocking CLN with 64 inputs can withstand five iterations of an SAT attack within 2×10^6 seconds. In contrast, a similarly sized blocking network can only resist the attack for about 17 s. Even a much larger blocking network with 512 inputs (16 times the area) can only withstand six iterations of an SAT attack within the same time frame.

6.3.1 Security Analysis

To implement Full-Lock, random insertion is used to add programmable logic and routing blocks (PLRs), which can create cycles in the design. CycSAT is used instead of SAT to handle potential cycles in locked circuits. Additionally, AppSAT is used to assess the resilience of circuits against approximate-based attacks by extracting the approximate key and corresponding error rate. It has been observed that including three PLRs, each containing 32×32 CLNs, makes locked circuits resistant to SAT for all circuits. However, using smaller PLRs can compromise CycSAT for each benchmark circuit.

To demonstrate the SAT-hardness of PLRs, the smallest size and the minimum number of PLRs needed to achieve resilience against SAT. Notably, Full-Lock consistently requires fewer and smaller PLRs compared to Cross-Lock [11]. For instance, in the *apex*4 circuit, only two PLRs with a 32×32 CLN and an additional PLR with an 8×8 CLN are sufficient to resist SAT within a 2×10^6 second timeout. Conversely, Cross-Lock [11] necessitates the insertion of 113 2×36 crossbars to achieve SAT resilience in the same circuit.

The average clauses-to-variables ratio for various LL schemes during de-obfuscation was calculated using MiniSAT to demonstrate that PLRs are instances that significantly increase the number (M) and computational complexity ($T_{\text{Avg DPLL}}$) of DPLL calls in each SAT iteration. In Fig. 6.4, it is evident that the clauses-to-variables ratio is 3.77 in Full-Lock, while it is much lower for all other methods. LUT-Lock and Cross-Lock in [11] exhibited

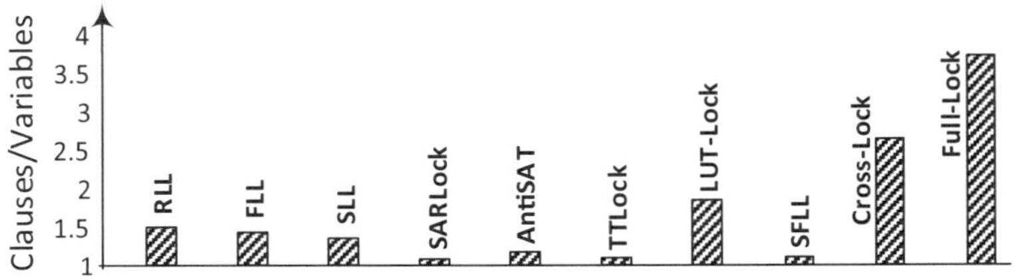

Fig. 6.4 Average clauses to variables ratio for RLL [1], FLL [12], SLL [13], SARLock [14], AntiSAT [15], TTLock [16], LUT-Lock [4], cross-lock [17] and full-lock [7]

higher clauses-to-variables ratios across all logic locking schemes. In the case of LUT-Lock, the utilization of key-programmable LUTs resulted in a Mux-based CNF translation. Still, due to the absence of back-to-back connections, the resulting Mux tree had low depth, consequently reducing the value of this ratio. On the other hand, Cross-Lock [11] showcased a closely matched clauses-to-variables ratio as it employed interconnect locking with a tree of Mux. Nonetheless, Full-Lock stood out with an almost 4 (3.77) ratio.

6.3.1.1 Security Against Removal Attack

Cross-lock [17] utilized high-density cone-based selection strategies, such as k-cut and wire-cut, to reduce the risk of removal attacks by limiting the selection of wires for crossbar insertion. However, the logic of the gates in Cross-lock [17] leading each CLN can be negated, making it possible to remove CLNs and determine the functionality of LUTs. On the other hand, it was noted that Full-Lock is not vulnerable to removal attacks and ensures correct functionality.

6.3.1.2 Security Against Algebraic Attack

An algebraic attack is a cryptanalysis method that exploits the mathematical structure of a cryptographic algorithm to uncover secret information, such as encryption keys [18]. This type of attack is effective against cryptosystems represented using algebraic equations. It involves translating the cryptographic algorithm into a mathematical form to work with a system of equations that describe the entire cryptographic process.

A CLN can be expressed as an affine transformation function of the data input X, of the form $y = A \cdot X + B$, where A is an $N \times N$ matrix and B is an $N \times 1$ vector, with all elements dependent on the key input. Although recovering A and B is not equivalent to finding the key input, it may enable the successful de-obfuscation of CLN. Since Full-Lock replaces the preceding gates of selected wires with LUTs, it cannot be transformed into an affine function. Thus, it is safe against SAT-based algebraic attacks.

6.4 Silicon Validation of LUT-Based Obfuscation

The majority of techniques present results from logic synthesis but do not cover chip fabrication. The first chip fabrication focused on LUT-based obfuscation is presented in [19]. The design flow differs from traditional physical design due to the addition of obfuscation primitives. This approach explores LUT-based obfuscation, which minimally changes the design flow and is resilient to SAT attacks. The logic synthesis stage also includes an iterative security-driven flow incorporating a security analyzer block. Gate selection, replacement, and security validation processes are repeated until security constraints are met. The gate selection process can be adjusted at this stage to avoid critical paths.

The gates are replaced with the preceding LUTs, which forward the proper inputs to the logical LUT, allowing n-input logic gates to be mapped to any m-input LUT by programming configuration bits (CBs) and input configuration bits. CB management is critical and must be protected from attackers. This can be achieved by storing them protected and loading them externally via the scan chain or using nonvolatile emerging devices as a secure and re-programmable solution [20, 21].

After adding the LUT-based obfuscation primitive, SAT resiliency is verified using the security analyzer block. An ML algorithm predicts the non-linear SAT runtime of an obfuscated IC, but representing the obfuscated netlist in a structured way that suits an ML model is challenging. To validate the resiliency of obfuscated IP against SAT attacks, the security analyzer block, models the converted CNF of an obfuscated circuit as an undirected and signed bipartite graph. The CNF bipartite graph comprehensively represents the CNF without losing information. Its multi-order version captures additional meanings of a CNF and allows for studying the effect of previous stages on a given gate (Fig. 6.5).

The presentation of a novel graph encoder layer has facilitated ML techniques for varying-size CNF bipartite graphs. The unique challenges these graphs pose are effectively addressed by leveraging the energy of the Restricted Boltzmann Machines (RBM). Each part of the energy function is comprehensively explained, including how interaction and individual literal and clause energies are computed.

After completing the analysis, the design flow can be executed traditionally. All the other stages, such as logic synthesis, gate-level design verification, placement and routing, and physical verification, are in place to achieve the final layout.

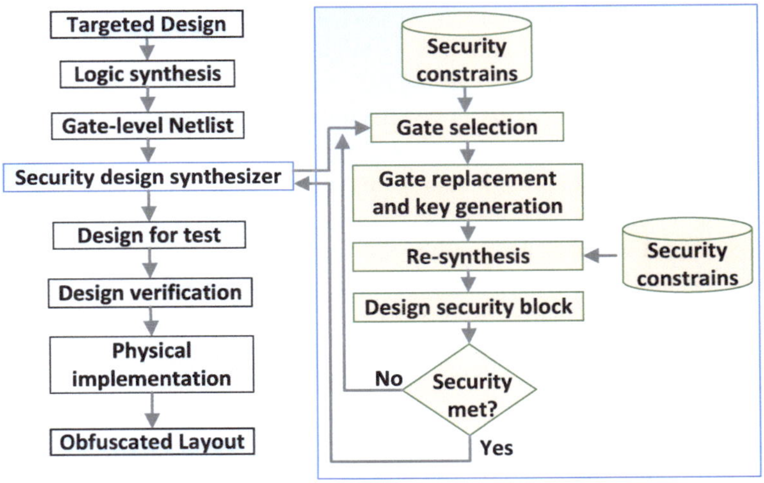

Fig. 6.5 Security oriented design flow for LUT-based obfuscation [19]

6.4.1 Security Analysis

For the security analysis, the dataset, consisting of around 21,000 obfuscated instances, was used to train a model. The obfuscated circuit was represented as an adjacency matrix, and the CNF bipartite graph was derived from this representation. Multiple-order information was extracted from the CNF bipartite graph, and an energy-based kernel was used to model the dynamic-size data.

The dynamic size of the CNF was handled using the energy concept from RBMs, and a distribution kernel was applied to model runtime variance for different instances. The model was evaluated using Pearson and Spearman coefficients and Mean Squared Error (MSE) scores. It was found that the model was effective in predicting the trend of runtime with low prediction error and did not predict the key. The model was used to predict the runtime for RISC-V benchmarks, and the obfuscation methodology used a LUT of size 8.

6.4.2 Chip Validation

Post-fabrication validation must be carried out to verify the functionality and security of the obfuscated design. The goal is to validate the functionality of the obfuscated design using data obtained after fabrication, as opposed to relying solely on simulated results. The functionality test ensures that the design operates correctly when the correct key is used and enters an obfuscated state when the wrong key is applied.

Two test chips were fabricated using TSMC 65nm technology to test the LUT-based obfuscation method. The first chip contains popular encryption engines (AES, DES, and SHA-3), a custom ALU, and other benchmarks. The second chip contains an obfuscated 32-bit RISC-V microprocessor core and its original non-obfuscated version. Both chips include various benchmarks to evaluate the LUT-based obfuscation method, demonstrating its scalability and performance impact at different obfuscation levels.

The key lengths for low, medium, and high obfuscated designs are 288, 576, and 3168 bits, respectively. Different key storage methods are used for the various obfuscated designs. For low and medium-obfuscated designs, the key bits are stored in non-volatile e-fuse, while for high-obfuscated designs, the key bits are stored in volatile SRAM registers. EFuse was chosen because it is the only embedded non-volatile memory available for the target fabrication technology. Additionally, SRAM storage was examined as a low-cost option readily available in standard CMOS, although it requires external and secure non-volatile key storage.

For chip validation, the designs are tested using test vectors generated from each design's original test bench. To ensure that the test bench properly stimulates the obfuscated cells, incorrect keys are also programmed for each benchmark, and failure is observed for improperly programmed designs. All three variants (low, medium, and high) result in SAT-timeout. These SAT-resilient benchmarks exhibit an average area overhead of 7% for low-obfuscated configurations, while medium and high obfuscation result in 14% and 262% overhead, respectively. Standby power scales dramatically as the security level increases, with increases of 33.93%, 45.93%, and 903.92% for low, medium, and high obfuscation, respectively. Conversely, the LUT-based obfuscation incurs only 0.03%, 3.53%, and 17.82% average dynamic power overheads for low, medium, and high obfuscations, respectively [19]. These results validate that standby power and area dominate the PPA cost as the LUT unit cell size increases. Timing results are not included as all designs maintained their original target frequencies of 200 MHz for the RISC-V core and 100 MHz for all other benchmarks. These are not very challenging frequencies and therefore it is hard to predict if the obfuscated designs can or cannot attain the same max frequency of operation.

6.5 Comparative Analysis and Discussions

LUT-based techniques generally provide reasonable security. In the subsequent discussions, we identify and address common pitfalls associated with these techniques.

6.5.1 Selection of LUTs

The technique described in [2] involves selecting gates and mapping them into LUTs. The selection criteria are based on controllability and observability, which are used as selection heuristics in logic networks and are expressed as don't-care conditions. However, this technique does not consider PPA and does not address the vulnerabilities exploited by the SAT attack and other attacks, such as structural attacks. It is worth noting that this technique was proposed before the SAT attack.

In contrast, the techniques in [4] focus on selecting LUT-based approaches, considering factors such as the fan-in cone of a minimum number of primary outputs, higher skew gates, gates with minimum fan-out, gates with less connection or impact on outputs, and back-to-back insertion of LUTs. When used in combination, all these selection methods aim to address SAT resiliency but do not take into account PPA when selecting the parts to be obfuscated.

On the other hand, Full-Lock [7] presents a completely SAT-targeted solution, which achieves SAT resiliency by integrating the CLN to obfuscate the routing. All their efforts are tailored to trap the DPLL algorithm and identify SAT hard locations for obfuscation. It is important to note that all the solutions are security-oriented with respect to SAT attacks.

6.5.2 Mitigation Strategies and Best Practices

All the techniques discussed lack a thorough security analysis. These approaches are only validated against the SAT attack instead of the novel attack. As shown in Fig. 6.6, the authors in [20] elaborated a rigorous security analysis of SAT, execution time versus the size of LUT,

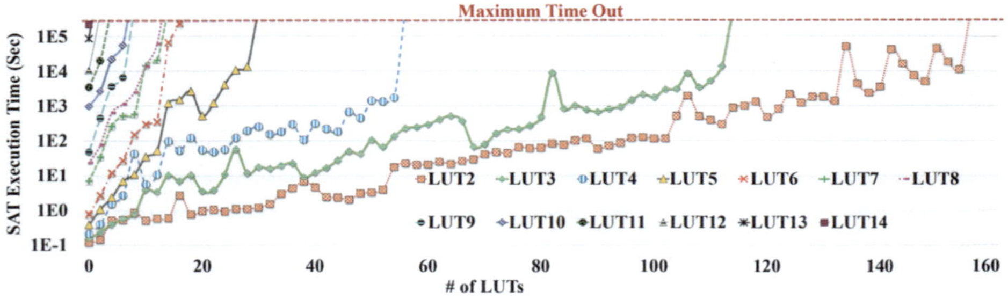

Fig. 6.6 Analysis of SAT attack execution time on *c*7552 circuit with varied numbers and sizes of LUTs [20]

and then execution time versus the key size for the *c*7552 circuit. It examines the impact of different LUT sizes and varying numbers of LUTs. The experiment reveals that using larger LUT sizes enhances SAT resiliency more effectively than obfuscating a higher number of gates with smaller LUT sizes. For instance, replacing a single gate with a 13-input LUT is adequate to achieve perfect resilience against SAT attacks. Downsizing the LUTs can help mitigate design overhead, but it comes at the cost of compromised security.

It is well understood that the size of the LUT is a significant and direct factor in the resilience of design. However, when considering PPA and physical implementation, the researches also proposed using NVM-based LUTs to address PPA overheads. We will describe the specifics in the upcoming chapters.

Besides the SAT attack, the attacks on SAT attack, such as structural attack [22–24], removal attack [25], and various other attacks applicable to LL can also be applied. In security and vulnerability analysis, it is essential to ensure that the methods and tools employed are rigorously validated against a diverse range of attack scenarios. Best practices dictate that approaches should be tested against both oracle-guided and oracle-less attacks. This gap highlights the need for more comprehensive validation procedures that encompass both oracle-guided and oracle-less scenarios. By doing so, security measures can be more robustly assessed, *this specific branch of obfuscation misses this analysis*.

6.5.3 Challenges in Chip Design

The majority of the solutions for LUT-based obfuscation have not been validated on silicon. In 2022, there was only one solution [19] that presented chip validation for this type of obfuscation. While synthesis results can provide a preliminary idea of the PPA, they do not offer insight into the actual security of the chip versus PPA overheads. The physical design implementation is a complex process with many intricate stages. Different placement strategies and optimization techniques can be applied, and during Place and Route, there may be instances where the targeted part of the design becomes the worst possible in terms of PPA, which many solutions overlook. There can also be cases where physical synthesis over optimizes the design and exposes obfuscation assets or diminishes their effectiveness.

On the other hand, several ReBO techniques introduce significant PPA overheads, which can be prohibitive in practical applications. These overheads often lead to a situation where the benefits of employing ReBO techniques are outweighed by the drawbacks, particularly in scenarios where efficiency and resource optimization are critical. As a result, the process of chip validation, which is crucial for ensuring the functionality of a design, may become uninteresting or even impractical. This limitation underscores the need for balanced approaches that can offer the advantages of ReBO techniques without compromising the PPA overheads that define the success of a chip design.

Now, we have presented the solutions and their security analysis in this chapter. In Chap. 7, we will discuss eFPGA redaction and its security analysis. It is noteworthy to remind that the LUT-based obfuscation is primarily a fine-grain approach, while the eFPGA redaction is a coarse-grain approach to obfuscation.

References

1. J. A. Roy, F. Koushanfar, and I. L. Markov, "EPIC: Ending piracy of integrated circuits," in *2008 Design, Automation and Test in Europe*, pp. 1069–1074, 2008.
2. A. Baumgarten, A. Tyagi, and J. Zambreno, "Preventing IC piracy using reconfigurable logic barriers," *IEEE Design Test of Computers*, vol. 27, no. 1, pp. 66–75, 2010.
3. Z. U. Abideen, S. Gokulanathan, M. J. Aljafar, and S. Pagliarini, "An overview of FPGA-inspired obfuscation techniques," *Association for Computing Machinery*, vol. 56, no. 12, December 2024.
4. H. Mardani Kamali, K. Zamiri Azar, K. Gaj, H. Homayoun, and A. Sasan, "Lut-lock: A novel lut-based logic obfuscation for fpga-bitstream and asic-hardware protection," in *2018 IEEE Computer Society Annual Symposium on VLSI (ISVLSI)*, pp. 405–410, 2018.
5. P. Subramanyan, S. Ray, and S. Malik, "Evaluating the security of logic encryption algorithms," in *2015 IEEE International Symposium on Hardware Oriented Security and Trust (HOST)*, pp. 137–143, 2015.
6. T. Winograd, H. Salmani, H. Mahmoodi, K. Gaj, and H. Homayoun, "Hybrid stt-cmos designs for reverse-engineering prevention," in *2016 53nd ACM/EDAC/IEEE Design Automation Conference (DAC)*, pp. 1–6, 2016.
7. H. M. Kamali, K. Z. Azar, H. Homayoun, and A. Sasan, "Full-lock: Hard distributions of sat instances for obfuscating circuits using fully configurable logic and routing blocks," in *Proceedings of the 56th Annual Design Automation Conference 2019*, DAC '19, 2019.
8. M. Soos, K. Nohl, and C. Castelluccia, "Extending sat solvers to cryptographic problems," in *Theory and Applications of Satisfiability Testing - SAT 2009* (O. Kullmann, ed.), (Berlin, Heidelberg), pp. 244–257, Springer Berlin Heidelberg, 2009.
9. H. Stone, "Parallel processing with the perfect shuffle," *IEEE Transactions on Computers*, vol. C-20, no. 2, pp. 153–161, 1971.
10. D.-J. Shyy and C.-T. Lea, "Log/sub 2/ (n, m, p) strictly nonblocking networks," *IEEE Transactions on Communications*, vol. 39, no. 10, pp. 1502–1510, 1991.
11. K. Shamsi, M. Li, D. Z. Pan, and Y. Jin, "Cross-lock: Dense layout-level interconnect locking using cross-bar architectures," in *Proceedings of the 2018 Great Lakes Symposium on VLSI,*

GLSVLSI '18, (New York, NY, USA), p. 147–152, Association for Computing Machinery, 2018.

12. J. Rajendran, H. Zhang, C. Zhang, G. S. Rose, Y. Pino, O. Sinanoglu, and R. Karri, "Fault analysis-based logic encryption," *IEEE Transactions on Computers*, vol. 64, no. 2, pp. 410–424, 2015.
13. M. Yasin, A. Sengupta, M. T. Nabeel, M. Ashraf, J. J. Rajendran, and O. Sinanoglu, "Provably-secure logic locking: From theory to practice," in *Proceedings of the 2017 ACM SIGSAC Conference on Computer and Communications Security*, CCS '17, p. 1601–1618, Association for Computing Machinery, 2017.
14. M. Yasin, B. Mazumdar, J. J. V. Rajendran, and O. Sinanoglu, "Sarlock: Sat attack resistant logic locking," in *2016 IEEE International Symposium on Hardware Oriented Security and Trust (HOST)*, pp. 236–241, 2016.
15. Y. Xie and A. Srivastava, "Anti-sat: Mitigating sat attack on logic locking," *IEEE Transactions on Computer-Aided Design of Integrated Circuits and Systems*, vol. 38, no. 2, pp. 199–207, 2019.
16. M. Yasin, B. Mazumdar, J. J. V. Rajendran, and O. Sinanoglu, "Ttlock: Tenacious and traceless logic locking," in *2017 IEEE International Symposium on Hardware Oriented Security and Trust (HOST)*, pp. 166–166, 2017.
17. K. Shamsi, M. Li, D. Z. Pan, and Y. Jin, "Cross-lock: Dense layout-level interconnect locking using cross-bar architectures," in *Proceedings of the 2018 Great Lakes Symposium on VLSI*, GLSVLSI '18, p. 147–152, 2018.
18. L. Chaoyun, "Algebraic cryptanalysis: A short introduction." https://www.esat.kuleuven.be/cosic/blog/algebraic-cryptanalysis/, 2024. Accessed: August 16, 2024.
19. G. Kolhe, T. Sheaves, K. I. Gubbi, T. Kadale, S. Rafatirad, S. M. PD, A. Sasan, H. Mahmoodi, and H. Homayoun, "Silicon validation of lut-based logic-locked ip cores," in *Proceedings of the 59th ACM/IEEE Design Automation Conference*, DAC '22, p. 1189–1194, Association for Computing Machinery, 2022.
20. G. Kolhe, H. M. Kamali, M. Naicker, T. D. Sheaves, H. Mahmoodi, P. D. Sai Manoj, H. Homayoun, S. Rafatirad, and A. Sasan, "Security and complexity analysis of LUT-based obfuscation: From blueprint to reality," in *2019 IEEE/ACM International Conference on Computer-Aided Design (ICCAD)*, pp. 1–8, 2019.
21. G. Kolhe, S. Salehi, T. D. Sheaves, H. Homayoun, S. Rafatirad, M. P. Sai, and A. Sasan, "Securing hardware via dynamic obfuscation utilizing reconfigurable interconnect and logic blocks," in *2021 58th ACM/IEEE Design Automation Conference (DAC)*, pp. 229–234, IEEE, 2021.
22. P. Chakraborty, J. Cruz, A. Alaql, and S. Bhunia, "SAIL: Analyzing structural artifacts of logic locking using machine learning," *IEEE Transactions on Information Forensics and Security*, pp. 1–1, 2021.
23. A. Alaql, M. M. Rahman, and S. Bhunia, "SCOPE: Synthesis-based constant propagation attack on logic locking," *IEEE Transactions on Very Large Scale Integration (VLSI) Systems*, vol. 29, no. 8, pp. 1529–1542, 2021.
24. A. Alaql, D. Forte, and S. Bhunia, "Sweep to the secret: A constant propagation attack on logic locking," in *2019 Asian Hardware Oriented Security and Trust Symposium (AsianHOST)*, pp. 1–6, 2019.
25. M. El Massad, S. Garg, and M. V. Tripunitara, "The SAT attack on IC camouflaging: Impact and potential countermeasures," *IEEE Transactions on Computer-Aided Design of Integrated Circuits and Systems*, vol. 39, no. 8, pp. 1577–1590, 2020.

eFPGA Redaction

7

7.1 eFPGA-Based Redaction Leveraging C/C++/Soft IP

Previously, in Chap. 6, we discussed LUT-based obfuscation, which relies on fine-grained reconfigurable elements. Now, our focus shifts to a coarse-grain approach described in [1] and named "design extraction onto an embedded FPGA (DEEF)".

At the HLS level, priority is given to the design [1], and it is suggested that parts of an ASIC design be selectively extracted to map it into an eFPGA in order to obfuscate the design's functionality. This obfuscation method involves the redaction of parts of the design, and then the bitstream is programmed to the fabricated silicon to restore the functionality of the design. FPGA-style LUTs are used to implement the redacted part of the design, providing high levels of security.

In the *first step*, the obfuscation method involves a crucial library pre-characterization step based on the targeted technology to obtain accurate area and delay information after HLS. The RTL description of an algorithm's untimed behavioral description is transformed through HLS, and resource allocation, scheduling, and binding are involved. Commercial HLS tools provide library characterizers to extract area and delay information for all operations in the target technology/FPGA design. The method continues by selectively extracting different portions of the behavioral description to analyze their effect on the circuit's area, delay, and security. This library pre-characterization step only needs to be performed once unless there is a change in the target technology.

In the *second step*, the portion to be obfuscated at the behavioral level is selected so the designer knows exactly what part of the design is obfuscated. This is not straightforward at the gate-netlist level, where the designer might lose track of what is obfuscated. The selective extraction method is composed of *three main parts*, as highlighted in Fig. 7.1 and Algorithm 7.1 lines 4–31.

Z. U. Abideen and S. Pagliarini, *Reconfigurable Obfuscation Techniques for the IC Supply Chain*, Synthesis Lectures on Digital Circuits & Systems,
https://doi.org/10.1007/978-3-031-77509-3_7

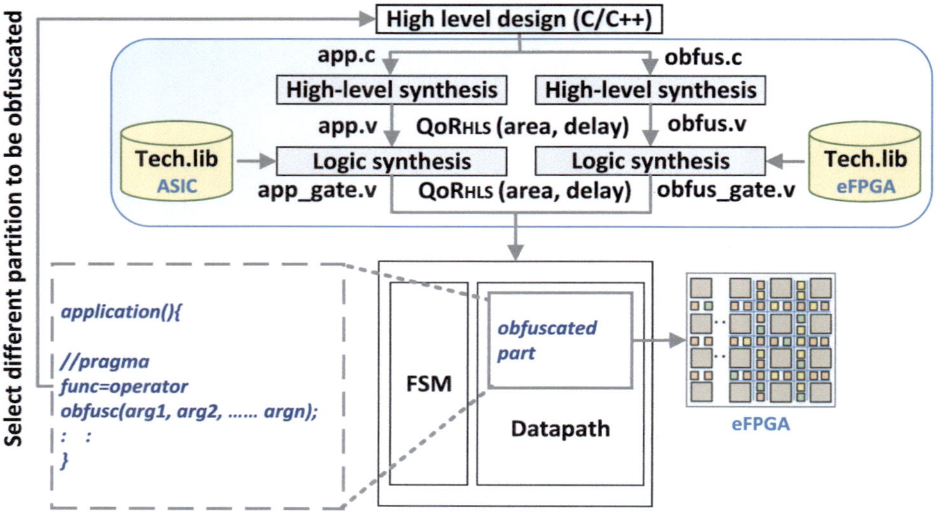

Fig. 7.1 Design obfuscation design methodology starting from a behavioral description for HLS [1]

The *first part* is function encapsulation involves selecting code lines and encapsulating them into a functional operator. HLS tools use pragmas to control array, loop, and function synthesis. Function synthesis enables encapsulating functions as functional units and controlling parallelism in the circuit using a constraint file.

This part also performs pragma annotation, which seeks to discover all mappable lines that qualify for obfuscation, whereas mappable implies a line that can be isolated and mapped as a functional operator. The pragma is annotated to all mappable lines by analyzing the code syntax into four types, such as operation line containing an assignment instruction, 'for' loop, 'if' or 'if-else' condition, and pragmas internal to an 'if' or 'if-else' condition. Lastly, type 4 encompasses all other syntaxes that do not qualify for mapping to the eFPGA. After each line is classified into one of the above four types, then pragma types are annotated.

When using the functional operator option to map the behavioral description to the eFPGA, two separate files are generated, as shown in Fig. 7.1: one containing the main application functionality and another containing the part to be obfuscated. These files can be synthesized separately. The main functionality uses the ASIC technology library, while the encapsulated portion uses the eFPGA technology library.

In the *second part*, the custom tool produces two RTL descriptions: one for the main module (`application.v`) and one for the encapsulated module (`obfus.v`). The security cost function S_i assesses the additional security achieved by mapping specific code portions to the eFPGA. The process is extended to include full logic synthesis after HLS, resulting in a more precise report (`QoR_HLS`) and gate netlists for both modules. Each line of C code is systematically evaluated to determine if it can be obfuscated and re-synthesized. If the area overhead exceeds a specified value, that line is not obfuscated.

Algorithm 7.1: Selective behavioral description extraction for obfuscation in eFPGA

Data: $\{C_{\text{orig}}, \text{Lib}_{\text{ASIC}}, \text{Lib}_{\text{eFPGA}}, f_{\text{target}}\}$
Input: C_{orig}: Behavioral description to be obfuscated
ASIC_db: ASIC technology library (db)
eFPGA_db: eFPGA technology library (db)
Result: $D_{\text{obfus\_opt}}$: Pareto-optimal $D_{\text{obfus\_merged}}$

1 /* **Pre-Characterization : Initialization** */
2 $HLS_{\text{TechLibASIC}} = \text{gen\_techlib(ASIC\_db)}$
3 $HLS_{\text{TechLib\_eFPGA}} = \text{gen\_techlib(eFPGA\_db)}$
4 /* **Selective Extraction: Part 1 Function encapsulation** */
5 Call Algorithm 1 (Function encapsulation)
6 /* **produces a set** $\{C_{\text{obfus\_single}}\}$ */
7 /* **Selective Extraction: Part 2 Individual Extraction** */
8 **for** *each* C_{obfus_single} **do**
9     $D_{\text{eFPGA}} = \text{hls\_eFPGA}(C_{\text{eFPGA}}, f_{\text{target}}, HLS_{\text{TechLib\_eFPGA}})$
10     $D_{\text{ASIC}} = \text{hls\_asic}(C_{\text{ASIC}}, f_{\text{target}}, HLS_{\text{TechLibASIC}})$
11     $D_{\text{obfus\_single}} = \{D_{\text{eFPGA}}, D_{\text{ASIC}}\}$
12   **if** $A_{total} > A_{max}$ **then**
13   discard $D_{\text{obfus\_single}}$
14 /* **Selective Extraction: Part 3 Results Merging** */
15 **for** *each* C_{obfus_single} **do**
16     $C_{\text{obfus\_merged}} = C_{\text{obfus\_single}}$
17   **for** *each subsequent* C_{obfus_single} **do**
18   /* **Merging** */
19         $C_{\text{obfus\_merged}} = C_{\text{obfus\_single}} \cup C_{\text{obfus\_merged}}$
20         $D_{\text{eFPGA\_merged}} = \text{hls\_eFPGA}(C_{\text{eFPGA\_merged}}, f_{\text{target}}, HLS_{\text{TechLib\_eFPGA}})$
21         $D_{\text{ASIC\_merged}} = \text{hls\_asic}(C_{\text{ASIC\_merged}}, f_{\text{target}}, HLS_{\text{TechLibASIC}})$
22   **if** $A_{total} \leq A_{max}$ **then**
23             $D_{\text{obfus\_merged}} = \{D_{\text{eFPGA\_merged}}, D_{\text{ASIC\_merged}}\}$
24   **else**
25             $C_{\text{obfus\_merged}} -= C_{\text{obfus\_single}}$
26 $D_{\text{obfus\_opt}} = \text{pareto\_optimal}(D_{\text{obfus\_merged}})$

In the *third part*, the results obtained are merged, where each result is derived from obscuring a single line of C code. In this step, multiple lines of C code are considered for potential simultaneous obfuscation. Each valid single obfuscated line is collected, and additional subsequent obfuscated lines of C code are searched for merging. Once the area constraint is violated, the merging stops and the process begins again with the next valid single line e (lines 19–30 in Algorithm 7.1). Each valid line is merged with subsequent lines until the area constraints are violated, which may result in considerable overlap.

7.1.1 Security Analysis

For the security evaluation, a security metric was introduced within the context of HLS [1], emphasizing the significance of obfuscation in terms of the number of operators assigned to an eFPGA block. Another crucial aspect that was highlighted was the implementation of a larger number of cells in the eFPGA, as it significantly complicates any potential SAT-based or brute-force attack. A security cost function for each obfuscated design was proposed, which is calculated by multiplying the number of cells by the number of operators (e.g., arithmetic or logical operations).[1]

$$S_i = \text{cell}_i \times \text{op}_i \tag{7.1}$$

Another possible attack method, besides a SAT-based attack, is a brute-force attack. This attack assumes the attacker is familiar with the eFPGA and can generate a bitstream. However, it lacks information on which bitstream configurations result in a valid circuit containing configured logic gates and their interconnections. The defined cost function is especially tied to the HLS-based eFPGA redaction.

7.2 eFPGA-Based Redaction Leveraging Firmware IP

Previously, we have stated the ReBO techniques, which leverage the HLS flow and soft IP. Now, we will illustrate the eFPGA redaction based on Firm IP named "Exploring eFPGA-based Redaction for IP (eRIP)" [2]. This approach exploits an open-source FPGA design flow to produce different eFPGA configurations, depending on the module to be redacted, and assesses the impact on a range of open-source IPs. eRIP focuses on factors contributing to the security provided by eFPGA-based redaction and the overall security of eFPGA fabrics.

To model a complete eFPGA fabric, auto-generated Verilog netlists can be fed into established ASIC design tool suites, especially Place & Route (P&R) tools, for generating layouts. In addition, self-testing Verilog testbenches can be automatically generated to ease pre- and post-layout verification. These Verilog testbenches validate the correctness of an eFPGA by simulating a complete process in practice, including bitstream downloading and eFPGA operation. The ability to create custom, small fabrics can provide a better fit for redaction compared with off-the-shelf eFPGA IP [3].

The eRIP flow involves the hierarchical design of IPs at the RTL and modification at the gate-level. To prepare the redaction fabric, the selected module is run through the OpenFPGA framework [4], which selects and generates the smallest eFPGA fabric configuration given an architecture definition. The generated fabric is simulated to verify that the intended functionality is correct, and if so, the synthesizable Verilog netlist is taken through a physical

[1] These operators are fundamental computational elements, such as addition, subtraction, and multiplication.

design flow that comprises synthesis, followed by floorplanning, placement, and routing. In contrast to related work [3], the eFPGA fabric is treated as a macro. After integrating this macro with the rest of the design, the IP, as a whole, is put through the design flow, resulting in a finalized layout.

When one redacts an IP, the selected module (the "redaction module") needs to fit into the eFPGA fabric; designers need to be aware of the resources available in a particular fabric size, especially if one were to adopt an HLS-based "top-down" approach [1]. The alternative approach is to find a fabric size that matches the requirements of the "designer-directed" redaction choice [3]. However, the minimum fabric size is driven by different factors of the redaction module. The interface of the module (number of inputs and outputs) will affect the number of I/O tiles required, while the number of state elements (registers/flip-flops) will affect the number of CLBs. Either factor can dominate the final eFPGA size, causing a sub-par use of the fabric used for redaction.

The selected redaction module could be on the critical path. FPGAs typically have longer delays than ASIC designs due to the abundance of available gates for logic and routing. As a result, the redacted portion in the eFPGA may be slower than in the ASIC. The designer must consider the impact on the IP's overall timing characteristics, including the redacted portion's effect, to avoid compromising the targeted performance.

Introducing an eFPGA fabric will considerably affect the area, particularly as the number of CLBs and I/O tiles increases non-linearly with each increase in the square eFPGA fabric's dimensions. This places another constraint on the design portion selected for the redaction. This redaction choice requires a fabric encompassing too much area in the context of the whole IP, which could be too impractical. In a related vein, the module selected for redaction could have numerous instances in the IP; the designer could create a larger fabric to redact several instances, instantiate multiple eFPGAs, or possibly choose to redact only one.

7.2.1 Security Analysis

The authors began with the SAT attack to clearly understand the attained security. SAT solvers encounter issues with combinational loops in eFPGAs, resulting in unstable results or repetitive return of input patterns. These loops arise from the reconfigurable routing network, adding complexity to the SAT formulation. To address this, pre-processing the netlist to add constraints is necessary. Solutions like BeSAT [5] and CycSAT [6] have limitations in resolving hard combinational loops in cyclic designs. IcySAT offers a practical loop-breaking alternative by identifying a subset of feedback nets, which, when removed, render the netlist acyclic.

Square eFPGA fabrics of different configurations were categorized based on the number of feedback nodes to be broken (Unroll Factor) and the clause size of the eFPGA netlist. The size of the IcySAT unrolled netlist is determined by the product of the unroll factor and the number of clauses required to represent the original circuit, both of which contribute to the complexity of the SAT attack. The attack increases exponentially as the eFPGA fabric increases in size. An average of 534 seconds was taken to complete the attack on the 3×3 fabric. Attempts to attack 4×4 and 5×5 fabrics were unsuccessful within 48 h, indicating that at least a 4×4 fabric should be selected for redaction.

In the analysis of eFPGA bitstreams, three main types of configuration bits were found: routing, logic, and I/O configuration bits. It was revealed through a partial SAT attack that recovering routing and logic bits is of similar complexity, while attacking I/O bits took longer. If routing and logic bits are assumed to be known, attacking I/O bits takes longer due to their low output corruptibility.

7.3 eFPGA-Based Redaction Leveraging Firmware IP and Open-souce Tools

The reconfigurability of eFPGA makes it a natural candidate for hardware obfuscation, but using eFPGA fabrics for fine-grained hardware redaction comes with its own challenges. We present another approach that also leverages the firm IP and open-source tools, which is "designer-directed fine-grained eFPGA insertion (DFGe)". DFGe explains the lack of CAD tool flows supporting combined logic synthesis, timing analysis, and optimization. The custom layout is time-consuming; therefore, using "soft" eFPGA-based designs specified in RTL and implementing the physical design with conventional standard cell-based CAD tool flow is more convenient.

The challenges mentioned earlier and achieving detailed hardware redaction are tackled using a soft eFPGA redaction flow. The approach involves leveraging multiple open-source tools to create soft eFPGA fabrics of different sizes and architectural parameters. An overview of a soft eFPGA redaction flow is displayed in Fig. 7.2. The designer starts with the Verilog description of the design, and the critical IP that needs to be redacted is marked. Then, the smallest eFPGA fabric required to map the redacted module onto an eFPGA is

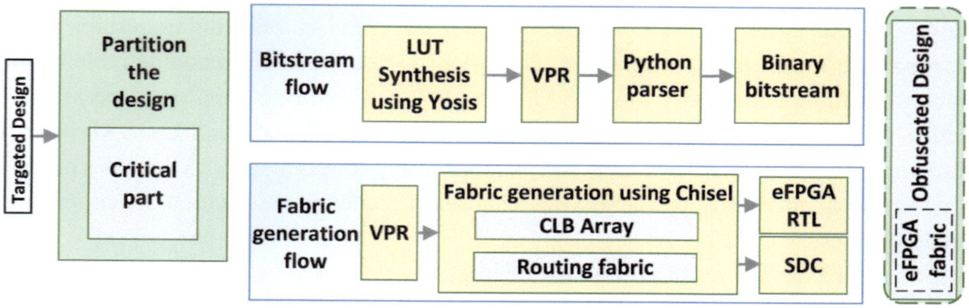

Fig. 7.2 Illustration of fine-grained hardware redaction using using open-source tools including Yosys, VTR/VPR, Chisel, and Cocotb [3]

automatically determined by the fabric generation flow. The corresponding eFPGA module is generated using a Chisel-based fabric generator, and the marked critical IP module is replaced with the eFPGA module to obtain the redacted design. The eFPGA module does not contain any design details of the security-critical IP, and the desired functionality can only be achieved by loading the eFPGA with the correct configuration bitstream. The static timing analysis (STA) constraints that can be used to synthesize and optimize the eFPGA fabric are also generated by the Chisel-based fabric generator. The redacted design, along with the generated STA constraints, are passed on to the standard ASIC physical design flow to generate the redacted layout.

The fabric generation flow is responsible for generating the redacted design with the eFPGA fabric, the bitstream required to map the critical IP onto the eFPGA fabric, and the STA constraints required for the physical design tools. The first step in the fabric generation flow is to synthesize the marked critical module into LUTs using the open-source Yosys synthesis tool [7]. The synthesized LUT netlist is then passed on to the VTR/VPR tool [8] to place and route the redacted module. During the place and route process, VPR identifies the smallest fabric size and routing channel width required to map the design onto an eFPGA fabric. The VPR output files are passed on to the fabric generator designed using the Chisel HDL [9]. The Chisel fabric generator outputs the eFPGA Verilog RTL along with the STA timing constraints that are necessary for downstream physical design tools. Bitstream generation is done by processing VPR output files through custom Python scripts. The resulting bitstream is used to verify the functionality of the redacted module using cocotb testbenches [10].

7.3.1 Security Analysis

The approach involved obfuscating two different designs: a RISCV CPU core and a GPS P-code generator. The process also included allocating security-critical modules and generic fabrics to assess the overhead and effectiveness of eFPGA redaction. Furthermore, the designs were fabricated using an industrial 22nm FinFET CMOS process.

The attack model assumes that the attacker has access to a fully scanned and unlocked design, as well as the netlist of the IP. Before launching a reverse-engineering attack, the attacker needs to overcome three challenges. Firstly, the eFPGA fabric needs to be isolated from the rest of the IP. Secondly, the attacker needs to obtain control and observation capabilities for the fabric's inputs, outputs, and flip-flops.

As previously explained, eFPGA fabrics contain numerous cycles due to their flexible interconnect network, which makes miter-based SAT attacks tend not to terminate. To facilitate our experiments, the CycSAT algorithm [11] was implemented to allow the attack to run to completion in cyclic circuits by adding constraints to the attack formulation. These constraints rule out any key that creates cycles in the design.

The resistance of eFPGA redaction to attacks was tested by conducting SAT attacks on a redacted RISC-V design with 8119 configuration bits. The SAT formulation included about 600 million clauses and 250 million variables. The attack consistently timed out. SAT attacks were conducted to explore the relationship between attack time and eFPGA configuration bit numbers. The number of unknown key bits was gradually increased, and the attack run times were plotted. These unknown key bits were evenly distributed between LUT and routing configuration bits. The results indicate that as the number of critical bits increases, the SAT attack runtime grows exponentially. All trials timed out when the number of key bits reached 1024, comprising 12.6% of the total configuration bits in the 4x4 tile eFPGA. In a non-timeout trial with 1024 key bits, the SAT attack took three days to complete, confirming the exponential growth of SAT runtime. In summary, the runtime of the SAT attack still follows an exponential trend for all designs.

7.4 Automating eFPGA Redaction Design Flow Leveraging Soft IP

Most eFPGA redaction approaches focus on enhancing security and minimizing the PPA overheads of the eFPGA. Still, addressing the EDA challenge of partitioning RTL modules between the eFPGA and ASIC and fabricating the appropriate eFPGA to realize these modules is crucial. The authors of [12] propose a comprehensive approach for tackling this EDA problem. Their proposed solution, named "automating eFPGA redaction design flow leveraging soft IP (ALICE)", involves a multi-step process that begins with identifying the modules to be implemented in the eFPGA and generating the corresponding soft eFPGAs. Subsequently, ALICE refines the solution iteratively by eliminating unsuitable modules, grouping the remaining ones to enable the creation of larger eFPGAs, and evaluating their hardware cost and security resilience to determine the optimal final implementation.

The process utilizes an openFPGA-based customization flow that begins with an XML specification of the fabric parameters and results in the creation of fabrication-ready eFPGA IP [2]. The modules to be redacted steer the customization of the eFPGA, enabling users to develop architectures that best fit the given design. The focus is on exploring an eFPGA architecture comprising CLBs constructed with four 4-input LUTs and prioritizing using fabrics over their generation and security assessment. Additionally, off-the-shelf fabrics can be integrated into the final chip. The redaction flow is depicted in Fig. 7.3. The implementation supports one or more eFPGAs with the same fabric architecture and a maximum number of used I/O pins.

ALICE can be split into three main phases: module filtering, cluster identification, and eFPGA selection. It commences with the RTL description of the design to be redacted in Verilog, along with a set of flow parameters stored in a custom YAML configuration file. These parameters include eFPGA fabric configurations (e.g., as specified in the OpenFPGA configuration file), the maximum number of eFPGAs to be instantiated, and the maximum

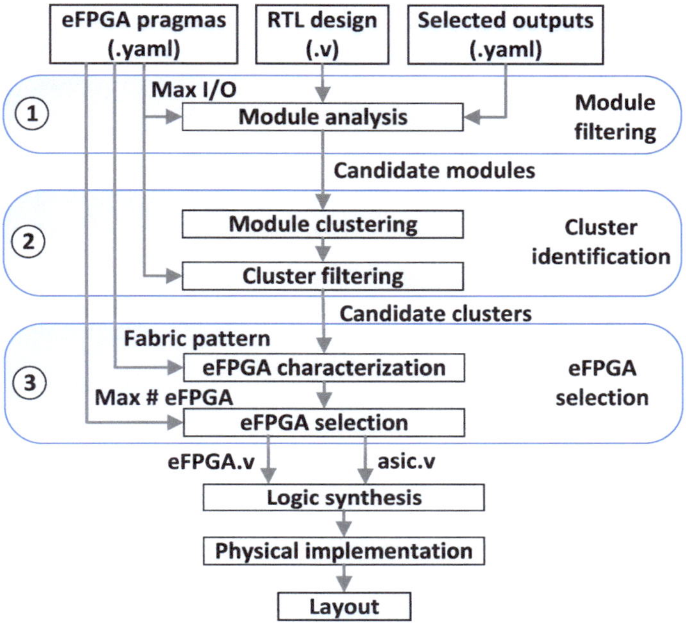

Fig. 7.3 Automatic flow for eFPGA-based redaction [12]

number of I/O for each. The number of I/O pins gives a rough indication of the type of eFPGA the designer intends to use.

During module filtering, ALICE analyzes the design to identify candidate redaction modules based on specific constraints. In the cluster identification phase, the candidate modules are grouped into clusters, and solutions that do not meet specific constraints are discarded. The resulting set of candidate clusters undergoes a flow process to create the corresponding eFPGA fabrics. Finally, an algorithm is applied to select the eFPGAs that meet given objectives, such as minimizing hardware overhead and maximizing security, without any overlap in redacted modules. The resulting redacted RTL description and the fabrics of the selected eFPGAs are then created, leading to the final system description that is ready for physical synthesis.

7.4.1 Security Analysis and Physical Implementation

ALICE focuses on the physical implementation and does not cover the security analysis of the proposed solution. ALICE was configured with two setups: cfg1 and cfg2. In cfg1, the maximum I/O pin count was set to 64 with a limit of two eFPGAs, while in cfg2, the maximum I/O pin count was 96 with a limit of one eFPGA. The eFPGA fabrics were

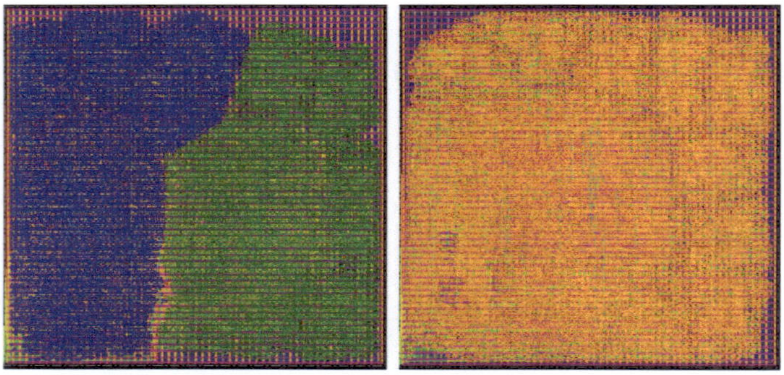

(a) cfg1: two 4×4 (52,629μm^2) (b) cfg2: one 5×5 (54,512μm^2

Fig. 7.4 Physical layouts of two GCD solutions featuring varying numbers of eFPGAs. [12]

implemented using the OpenFPGA flow, and the designs were implemented using Cadence Genus and Cadence Innovus. With the exception of one benchmark, at least one feasible solution for the given eFPGA parameters could be found. Interesting results were obtained from benchmarks like DES3 and the greatest common divisor (GCD) since these designs have more instances than the other benchmarks. ALICE was able to identify several candidate clusters in both configurations (over 200 for DES3 and at least 19 for GCD). Modules with a wide variance in the number of I/O pins were found in both GCD and DES3. Multi-module redaction clusters were possibly created by combining modules with low and high numbers of I/O pins, but clusters with many I/O pins became invalid. The use of more eFPGAs (cfg1) led to a significant growth in the number of possible cluster combinations and solutions. Solutions with smaller eFPGAs (two 4×4 for GCD) or one larger eFPGA (5×5 for GCD) were found. For DES3, the second implementation with a 14×14 eFPGA can be utilized as it redacts many more modules than the first case. The layouts of the physical designs for GCD are depicted in Fig. 7.4. This test case is small, so the chip is occupied by the FPGAs. Both solutions are equivalent from an area viewpoint and have almost the same number of redacted modules in the case of GCD.

7.5 Comparative Analysis and Discussions

In this chapter, we discussed many approaches towards eFPGA redaction, each with its way of obfuscating. These methods use soft IP, firm IP, or hard IP for eFPGA and automate the design process. The use of eFPGA utilizing HLS and soft IP was explored by the authors in [1], but the availability of a ready-made eFPGA hard macro IP to obfuscate the designs was assumed. While eFPGA hard macro IPs can be purchased from commercial vendors such as Achronix, QuickLogic, and Menta, these are not suitable for *fine-grained obfuscation* because eFPGA hard macro is typically organized into very large blocks (1K–4K LUTs) and the eFPGA architecture parameters cannot be fine-tuned by the end user to match the redacted design [13–15]. The ability to rapidly and automatically generate eFPGA fabrics of different sizes and architectural parameters that match the redacted portions of the design is needed to enable fine-grained eFPGA redaction. Thus, most approaches involve partitioning the design at either a soft IP or firm IP level. eFPGA redaction typically offers reasonable security and is resistant to SAT attacks. In the following subsections, we will highlight and address common issues related to these methods.

7.5.1 Considerations with Mixed ASIC/eFPGA Physical Design Flow

A generic eFPGA IP has to offer satisfactory delay, power, and area metrics for any application mapped onto the fabric. Generic timing optimization is necessary for the paths between the logic blocks so that the logic synthesis is application-agnostic. This is not the case for an eFPGA that will be used to redact a *specific design*. This means that the paths within the fabric that will be exercised during operation are also known, and the CAD tools can use this information to optimize the fabric further. Although this optimization would minimize the overhead of redaction, the buffering and gate sizing efforts by the tools might make the optimized paths easier to determine and may help the attackers narrow down some of the routing configuration bits by analyzing structural hints. In other words, the loss of regularity may benefit an adversary.

Secondly, the eFPGA is treated as a macro and it is coarse-grain implementation. Strategic placement of the hard macro within the chip is crucial. This involves balancing the PPA trade-offs at the SoC level. The chosen redaction unit can possibly lie in the critical path. FPGA structures tend to have longer delays compared to full ASIC designs due to the general nature of the large pool of available gates for logic and routing. Thus, the redacted portion in the eFPGA will likely be slower compared to the same design implemented directly in the ASIC.

Ensuring that the hard macro meets power constraints is crucial. This includes both dynamic power (due to switching activity) and static power (due to leakage currents). Inefficient placement can increase power consumption due to longer interconnects and higher switching activities. Proper design of the Power Distribution Network (PDN), and of the floorplan as a whole, is essential to provide stable voltage levels across the macro. Issues like IR drop (voltage drop due to resistance in power lines) and electromigration (degradation due to high current density) must be managed.

The introduction of an eFPGA fabric will have significant implications on area, especially as the number of configurable logic blocks (CLBs) and input/output (I/O) tiles increases exponentially with each increase in the size of the square eFPGA fabric. This imposes a constraint on the selected design–choosing a fabric that occupies too much area, in the context of the entire IP, could be impractical.

7.5.2 Lack of Comprehensive Security Analysis

In many solutions, it is stated that the adversary needs to be able to distinguish between the eFPGA and ASIC components in their threat model. When it comes to eFPGA redaction, visually inspecting the layout can make it easy to differentiate the eFPGA and non-reconfigurable parts. This assumption is reasonable and widely accepted. Researchers often claim that eFPGA redaction is resistant to SAT attacks and that SAT cannot compromise security. This is also evident from the eFPGA redaction techniques discussed in this chapter. Currently, two common types of SAT analysis are "clause-to-variable ratio" and "bitstreams versus execution time". The solution "Full-Lock" discussed in the previous chapter, uses the aforementioned analysis to define its technique as SAT resistant [16]. No doubt, the SAT attack is a powerful attack against any obfuscation technique. However, it is important to note that the SAT attack could succeed on eFPGA redaction if enough computational power is available. In other words, making use of eFPGAs does not prevent the attack from being mounted. Furthermore, this type of obfuscation also lacks a security evaluation for various other types of attacks, including structural attacks, resynthesis attacks, and fault-based attacks. This necessitates the advancement of additional attacks that specifically focus on the nature of ReBO in order to assess its robustness [17].

7.5.3 Security of Bitstream

It is crucial to understand end-user security when implementing ReBO techniques that involve using and storing bitstreams. Physical emanations from the ReBO could be exploited by side-channel attacks, such as power analysis and electromagnetic analysis, to extract the bitstream. While side-channel attacks primarily target cryptography-capable circuits, they can also reveal vulnerabilities in other design implementations [18]. Secure access to the encrypted bitstream requires authentication for both the user and the bitstream itself.

So far, we have discussed various eFPGA redaction techniques of different classifications. Yet, all approaches discussed so far are created using traditional CMOS technology. In Chap. 8, we will discuss methods that utilize emerging technologies like STT-MTJ, SOT-MTJ for designing the LUT. We will also explore how these approaches differ from traditional SRAM-based LUTs. standard

References

1. B. Hu, T. Jingxiang, S. Mustafa, R. R. Gaurav, S. William, M. Yiorgos, C. S. Benjamin, and S. Carl, "Functional obfuscation of hardware accelerators through selective partial design extraction onto an embedded FPGA," in *Proceedings of the 2019 Great Lakes Symposium on VLSI*, p. 171–176, 2019.

2. J. Bhandari, A. K. Thalakkattu Moosa, B. Tan, C. Pilato, G. Gore, X. Tang, S. Temple, P.-E. Gaillardon, and R. Karri, "Exploring eFPGA-based redaction for IP protection," in *2021 IEEE/ACM International Conference On Computer Aided Design (ICCAD)*, pp. 1–9, 2021.
3. P. Mohan, O. Atli, J. Sweeney, O. Kibar, L. Pileggi, and K. Mai, "Hardware redaction via designer-directed fine-grained eFPGA insertion," in *2021 Design, Automation & Test in Europe Conference & Exhibition (DATE)*, pp. 1186–1191, IEEE, 2021.
4. X. Tang, E. Giacomin, B. Chauviere, A. Alacchi, and P.-E. Gaillardon, "Openfpga: An open-source framework for agile prototyping customizable fpgas," *IEEE Micro*, vol. 40, no. 4, pp. 41–48, 2020.
5. Y. Shen, Y. Li, A. Rezaei, S. Kong, D. Dlott, and H. Zhou, "Besat: behavioral sat-based attack on cyclic logic encryption," in *Proceedings of the 24th Asia and South Pacific Design Automation Conference*, ASPDAC '19, (New York, NY, USA), p. 657–662, Association for Computing Machinery, 2019.
6. H. Zhou, R. Jiang, and S. Kong, "Cycsat: Sat-based attack on cyclic logic encryptions," in *2017 IEEE/ACM International Conference on Computer-Aided Design (ICCAD)*, pp. 49–56, 2017.
7. W. Clifford, "Yosyshq - open source eda." https://github.com/YosysHQ, 2024. Accessed: July 24, 2024.
8. K. E. Murray, O. Petelin, S. Zhong, J. M. Wang, M. Eldafrawy, J.-P. Legault, E. Sha, A. G. Graham, J. Wu, M. J. P. Walker, H. Zeng, P. Patros, J. Luu, K. B. Kent, and V. Betz, "VTR 8: High-performance cad and customizable FPGA architecture modelling," *ACM Transactions on Reconfigurable Technology and Systems*, vol. 13, no. 2, 2020.
9. J. Bachrach, H. Vo, B. Richards, Y. Lee, A. Waterman, R. Avižienis, J. Wawrzynek, and K. Asanović, "Chisel: constructing hardware in a scala embedded language," in *Proceedings of the 49th Annual Design Automation Conference*, DAC '12, p. 1216–1225, 2012.
10. Cocotb, "Use cocotb to test and verify chip designs in python. productive, and with a smile." https://www.cocotb.org/, 2024. Accessed: August 16, 2024.
11. H. Zhou, R. Jiang, and S. Kong, "CycSAT: SAT-based attack on cyclic logic encryptions," in *2017 IEEE/ACM International Conference on Computer-Aided Design (ICCAD)*, pp. 49–56, 2017.
12. C. M. Tomajoli, L. Collini, J. Bhandari, A. K. T. Moosa, B. Tan, X. Tang, P.-E. Gaillardon, R. Karri, and C. Pilato, "ALICE: An automatic design flow for eFPGA redaction," in *Proceedings of the 59th ACM/IEEE Design Automation Conference*, p. 781–786, 2022.
13. Achronix Corp., "Speedcore embedded FPGA IP," last accessed on Apr 22, 2023. Available at: https://www.achronix.com/product/speedcore.
14. QuickLogic Corp., "efpga ip 2.0 – enabling mass customization with fpga technology," last accessed on Jan19, 2023. Available at: https://www.quicklogic.com/products/efpga/efpga-ip2/.
15. Menta, "Embedded FPGA IP," last accessed on Apr 2, 2022. Available at: https://www.menta-efpga.com/.
16. H. M. Kamali, K. Z. Azar, H. Homayoun, and A. Sasan, "Full-lock: Hard distributions of sat instances for obfuscating circuits using fully configurable logic and routing blocks," in *Proceedings of the 56th Annual Design Automation Conference 2019*, DAC '19, 2019.
17. Z. U. Abideen, S. Gokulanathan, M. J. Aljafar, and S. Pagliarini, "An overview of FPGA-inspired obfuscation techniques," *Association for Computing Machinery*, vol. 56, no. 12, December 2024.
18. B. Yang, K. Wu, and R. Karri, "Scan based side channel attack on dedicated hardware implementations of data encryption standard," in *2004 International Conferce on Test*, pp. 339–344, 2004.

ReBO Leveraging Emerging Technologies 8

8.1 STT-MTJ-Based LUT Implementation

Previously, in Chap. 7, we discussed eFPGA redaction, which relies on coarse-grained reconfigurable elements. Let us briefly touch on ReBO leveraging emerging technologies, where the reconfigurable element is now fine-grained and designed with a hybrid of CMOS and emerging technologies. The STT-MTJ-based obfuscation method involves stacking an STT device with multilayer sandwich structures that are commonly used in the industry. As shown in Fig. 8.1, these structures comprise multiple layers, including an oxide tunnel barrier, a free magnetic layer, and a pinned magnetic layer. The orientation of the free layer's magnetization can be changed from a parallel to an antiparallel state (P to AP) by either an external magnetic field or a spin-polarized current J_{read} passing through the junction. These states can then represent a logic-1 or a logic-0. It is worth noting that the MTJ itself, depicted in Fig. 8.1, does not take up space that would otherwise be used by standard cells, as it sits between two metal layers.

The switching mechanism occurs in-plane when the current density exceeds a critical value, J_c, which is as low as $8 \times 105\,A/cm^2$ in CoFeB/MgO/CoFeB stack structures [1]. Since the spin-MTJ device surface area is typically small (in the order of $100\,nm \times 100\,nm$ or smaller), the critical current is less than $100\,\mu A$ and can be generated by a typical CMOS current source. Additionally, J_c can be reduced by integrating perpendicular magnetic anisotropy (PMA) in the free layer of the MTJ so that stable magnetization points out-of-plane instead of within the same plane as the free layer [2].

When integrating MTJs, it is important to note that they require minimal die area, with the exception of the CMOS circuits and contacts used to connect MTJs to MOS transistors. However, it is worth considering that MTJs come with operational challenges, particularly related to asymmetry in write and read operations. This asymmetry results in differences in operation energy and delay, which in turn require a higher current for completing write

Z. U. Abideen and S. Pagliarini, *Reconfigurable Obfuscation Techniques for the IC Supply Chain*, Synthesis Lectures on Digital Circuits & Systems,
https://doi.org/10.1007/978-3-031-77509-3_8

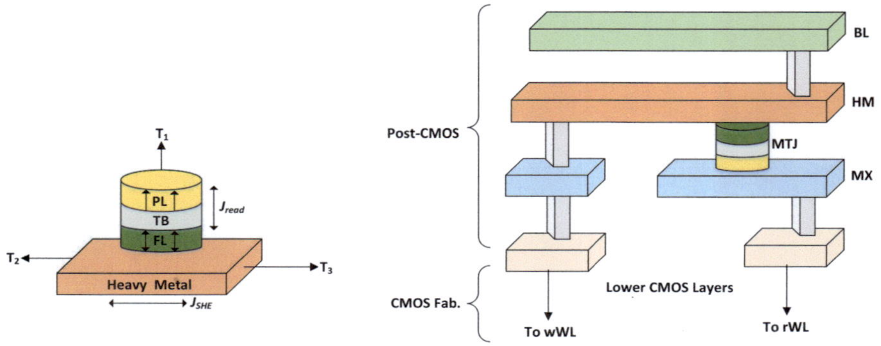

Fig. 8.1 Physical structure demonstration of SHE assisted MTJ switching mechanism [3]

operations. With this in mind, the authors in [4, 5] consider STT-MTJ-based LUT design for obfuscation. They propose STT-MTJ-based LUTs where the reconfigurable bits are stored in an MTJ, as the name implies. MTJ are inserted between metal layers, as shown in Fig. 8.1. The delayering process involved in the reverse engineering of IPs to retrieve the netlist results in the destruction of the structure of the MTJ device and, thus, will result in the loss of data stored in the MTJ. The MTJ, in this manner, serves as the tamper-proof memory to store the configuration of the LUT.

The custom component of the STT-MTJ-based LUT design leverages the standard cell-based ASIC design flow. The resulting designs are static due to the use of static logic style in implementing the ASIC standard cells. Consequently, the STT-MTJ-based-LUT design is constrained to a static-type interface for integration with the static ASIC standard cells. In the proposed STT-MTJ-based LUT structure, the path from the LUT inputs to the LUT output is represented by a Mux. The Mux of the LUT is a 2^n to 1 ($2^n : 1$) CMOS Mux implemented in a static style and can be expressed as synthesizable RTL code.

The storage of each configuration bit is achieved using an MTJ latch with scan chain programmability, as depicted in Fig. 8.2. This latch utilizes a pair of differentially programmed MTJs for non-volatile storage, a pre-charge sense amplifier for observing the state of the MTJs, and three write driver schemes for concurrently writing MTJs, with each MTJ receiving a full voltage swing to provide a higher write current. For successful write operations, the sense enable (SE) signal needs to be low, while the write enable (WE) signal must be low during sensing operations. To prevent conflicts between the pre-charge state of the sense amplifier (when $SE = 0$) and the state of the write driver outputs in the write mode, the path to V_{DD} is disconnected via the PMOS, which is controlled by the WE signal.

When using the MTJ latch, activating the dynamic latched sense amplifier with a low to high pulse during each power-up is essential. This converts the resistive state of the MTJs into volatile voltage states at the outputs (Q and QB). After this, the MTJs become read-only, and in active mode, the LUT read power and delay are determined by the static Mux. Unlike in dynamic STT-MTJ-based LUT, avoiding repetitive reading from the MTJs during

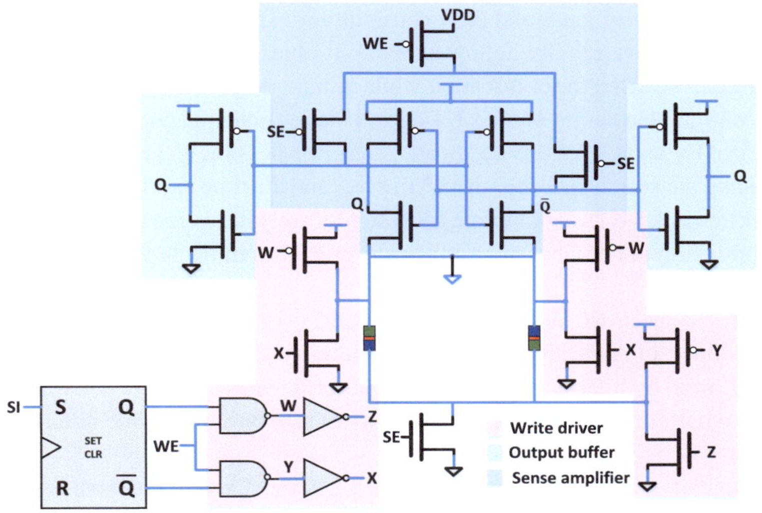

Fig. 8.2 MTJ latch with scan chain programming [5]

active mode reduces stress on the MTJs, ultimately improving their lifespan. Additionally, the scan chain used to program the LUT is separate from the one used in the design for traditional testing purposes.

After creating the STT-MTJ-based LUT, the next step is to substitute the gates with LUTs. The location of the gates selected for the obfuscation is one of the critical factors that define the strength of the obfuscation. The replacement strategy finds the gate for obfuscation based on heuristics. Replacement strategies affect SAT resiliency because the replacement strategy needs to be met to provide resiliency against the SAT attacks. The two most important conditions are (1) low corruptibility and (2) avoiding unintentionally correct key generation. The impact of each condition is compared using three strategies: random selection, low output corruptibility, and unintentional correct key generation avoidance.

The gates are selected for obfuscation in a random fashion using a random selection algorithm. This method is used as a baseline for comparison instead of the independent selection [6].

State-of-the-art SAT solvers use the conflict-driven clause learning (CDCL) algorithm to find solutions by searching for conflicting clauses. By comparing different outputs of the same netlist with different input patterns, the conflicting clause can be found easily if the Hamming distance between the outputs is high. An obfuscation strategy that affects multiple primary outputs when a wrong key input is applied results in a higher Hamming distance. This high corruption leads to faster detection of conflicting clauses and lower de-obfuscation time. Conversely, *low output corruptibility* makes it harder to sensitize the key input to the primary output. To achieve low output corruptibility, a breadth-first search on the graph is employed to create a dictionary of gates and their corruptibility. This allows for selecting gates with the lowest output corruptibility, followed by multi-objective optimization to maximize the number of gates selected for obfuscation while minimizing the output corruptibility.

The reconfigurable nature of the LUT allows for 2^{2^n} potential configurations, where n is the LUT size. By using LUTs to obfuscate gates, it is possible to generate additional valid keys, reducing the search space for the SAT solver and the time required for de-obfuscation. Using back-to-back LUTs to obfuscate gates significantly decreases the SAT solver's search space. Minimizing the number of valid keys is crucial to mitigate this scenario due to the LUTs' reconfigurability.

To directly connect two obfuscated gates, the *LC_NoGen* replacement strategy was proposed as outlined in Algorithm 8.5. This algorithm entails graph traversal and dictionary creation, followed by filtering to exclude back-to-back gates and those contributing to critical paths. An optimization process then maximizes gate coverage while minimizing output corruption.

The complexity of graph traversal with V vertices and E edges is denoted as $O(V + E)$. Integrating dictionary creation and filtering into the graph traversal process with minimal adjustments results in an overall complexity of $O(V + E)$. However, the optimization problem presents a combinatorial challenge. Although the task of finding the optimal set of gates for obfuscation is an NP-hard problem, a sub-optimal solution is approximated using off-the-shelf ILP solvers. The gate selection task, aimed at minimizing output corruptibility, is treated as a variant of the classic minimum vertex cover problem, where the objective is to select the minimal number of nodes to maximize coverage.

8.1.1 Results and Security Analysis

The experimental assessment utilized ISCAS'85 & ISCAS'89 benchmarks and applied an SMT-based attack [7]. The attack involves a primary SAT solver and additional theory solvers. Emphasis was placed on the runtime of the SAT attacks as a crucial metric for evaluating the security of the obfuscated circuit. Additionally, the execution time of the SMT solver was examined while varying the size of the MTJ-based LUT from 4 to 7 bits to showcase the impact of the LUT on security design. Furthermore, a runtime limit of 30 days (2,592,000 s) was set for the SMT attack to demonstrate time-out states.

The increase in power consumption and circuit area is noticeably observed when the size of encryption keys is increased. By limiting the number of key bits, the area, power, and sizes of LUTs used for encrypting data are indirectly limited. For the AES benchmark, varied key lengths are considered: 110, 160, 360, and 400 keys. These key lengths allow the use of LUTs of different sizes and determine the number of gates to be replaced. Increased resistance to de-obfuscation is achieved by using larger LUTs and a smaller quantity. Moreover, the added overhead is roughly the same when key lengths are equal. Additionally, security grows faster than the added overhead when larger LUTs are used in smaller quantities. The experiment shows that it is possible to maintain security while minimizing design overhead.

Algorithm 8.5: Avoiding Unintentionally Correct Key Generation (LC_NoGen) [5]

```
for each LCO in Logic_cones do
    // LCO: Logic Cone Output
    gate_list = BFS(LCO);
    // Get all gates in the logic cone
    for each gate in gate_list do
        gate.listLCOs = find_affected_LCOs(gate);
        // Find all the Logic Cone Output gates affected by the
           current gate
    end
end
for each gate in circuit do
    for each LCO in gate.listLCOs do
        tag_key(LCO);
        if isExist(tag_key(LCO)) then
            dictionary.add(gate);
        end
        else
            dictionary.add(gate);
            dictionary.addtag(tag_key(LCO));
        end
    end
end
CriticalPath = PrimeTime(Get_Critical_Path);
// Get list of gates that are on Critical Path using Synopsys
   PrimeTime
for each tag in dictionary do
    for each gate in tag do
        if isExist(Parent(gate) in tag) or isExist(gate in CriticalPath) then
            dictionary[tag].delete(gate);
            // Remove gates that are adjacent to each other to avoid
               back-to-back LUT replacement
            // Remove gates on critical path list
        end
    end
end
tag_key = Optimize();
// Find tag_key which has maximum gate coverage with lowest
   Output Corruption
for each gate in tag_key do
    Replace_LUT(gates, target_no);
    // Replace gates with LUT
end
```

8.2 Security-Driven Flow for STT-MTJ Based LUT Implementation

In the previous section, it was presented that STT technology could be used to construct resilient LUTs that are resistant to reverse engineering. It was learned that the STT-MTJ-based design possesses functionality similar to that of an FPGA, offering increased speed, reduced leakage power, and enhanced thermal stability. Unlike SRAM-based LUTs, non-volatile STT-MTJ-based LUTs eliminate the necessity for additional flash memory to store configuration bits for loading upon power-up. It is noteworthy that significantly higher write current is required by STT-MTJ-based LUTs compared to SRAM-based LUTs.

One notable advantage of STT-MTJ based LUTs is that their power and delay are independent of the logic they implement and the input data activity. Instead, these LUTs only rely on their fan-in (number of inputs). Additionally, due to their usage of fewer PMOS transistors, STT-MTJ based LUTs offer significant benefits, particularly in implementing logic gates.

The security-focused design process for STT-MTJ-based LUT is illustrated in Fig. 8.3 [6]. This approach seamlessly integrates with the standard IC design flow while prioritizing early-stage security to prevent design reverse engineering and minimize the impact on design constraints. The design flow begins with identifying design security requirements, STT technology library details, design constraints, and CMOS technology node information. Circuit implementation and logic synthesis follow this initial stage. Subsequently, the gate-level netlist obtained from the logic synthesis is passed on to the "CMOS gate selection and

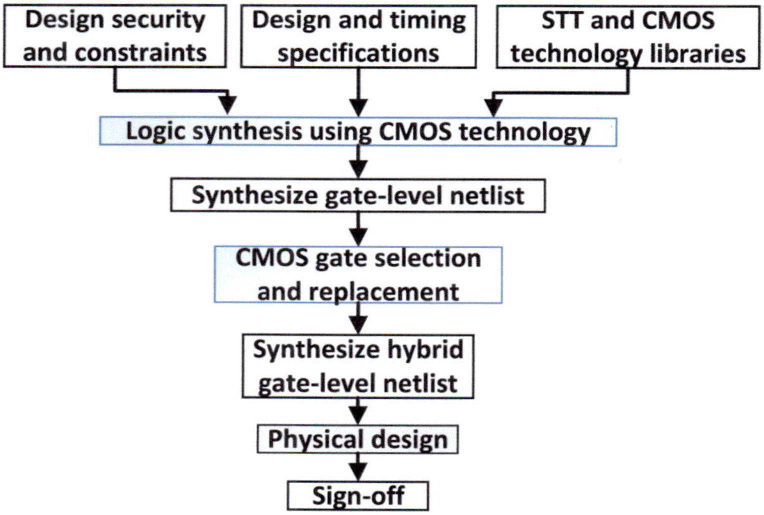

Fig. 8.3 Security-driven hybrid STT-CMOS design flow (adapted from [8])

replacement stage". At this point, the designer selects one of the proposed modes depending on the design security requirements.

In the first mode, gates are randomly selected and may not be connected. The selected mode replaces several CMOS gates in the synthesized gate-level netlist with equivalent STT-MTJ-based LUT implementations. While this approach provides some level of security, an attacker with adequate resources could still manipulate the functionality of specific components. Attackers can employ testing techniques to determine the functionality of missing gates and propagate their outputs to observation points. This also depends on the number of missing gates (M) and the depth of the circuit (D), where the depth is defined as the maximum number of FFs on a path from a primary input to a primary output of a circuit.

In the second mode, called dependent selection, gates to be mapped on STT-MTJ based LUTs are selected based on their accessibility to each other. This method adds complexity to determining the functionality of missing gates. All gates on the longest input/output path between a primary input and a primary output are replaced with STT-MTJ based LUTs, including timing paths that start and end with FFs.

The third mode is parametric-aware dependent selection. Instead of replacing all timing path gates with STT-MTJ based LUTs, only a select few gates are chosen for replacement. This approach minimizes the impact on timing requirements. Specifically, it targets the longest input/output paths and replaces gates with two or more inputs with STT-MTJ-based LUTs, increasing the effort required for reverse engineering. The selection is repeated if a timing violation occurs after the replacement. Furthermore, any unselected gates on the targeted timing path are saved in an unselected list (USL), and any gates connected to these unselected gates are replaced with STT-MTJ-based LUTs.

During the design implementation phase, the gates were selected and mapped to STT-MTJ-based LUTs in the circuit netlist. While STT-MTJ-based LUTs may initially have higher overhead than custom CMOS gates, this overhead decreases as circuit complexity increases. Upon obtaining the hybrid netlist, the design process advances to physical design, followed by the final sign-off. The security-aware hybrid STT-CMOS design flow incorporates design security requirements and reconfigurability from the early stages, making it highly resilient to reverse engineering attacks. Incorporating a circuit with missing gates prevents untrusted foundries from duplicating the design, as the design house or authorized vendors must ultimately configure each design. Moreover, the strategic selection and replacement of CMOS gates make it extremely challenging to identify missing gates within a reasonable timeframe.

8.2.1 Results and Security Analysis

ISCAS'89 benchmarks were used to assess security in the 90nm technology node using Synopsys Design Compiler as a synthesis tool. The assessment focused on gate selection and replacement, followed by the physical design to fabricate the design. It is important to note that the random selection of timing paths and gates resulted in slightly higher overhead for larger circuits in some cases, while slightly smaller circuits incurred lower overhead. For instance, there was a noticeable difference between the $s1288$ and $s1488$ benchmarks, as well as the $s1238$ and $s1196$ benchmarks.

In terms of security, a study was conducted to determine the number of clocks needed to identify missing gates in circuits using machine learning attacks. The findings revealed that even for small circuits, the number of test clocks required for the parametric-aware selection is incredibly high. For instance, the analysis of the $s38584$ benchmark demonstrated that introducing only 166 STT-MTJ-based LUTs using the parametric-aware selection technique would demand approximately 6.07×10^{219} test clocks to establish their functionality, with just a 5.13% increase in power consumption, a 1.56% increase in area, and no performance loss.

8.3 STT-MTJ Based LUT Implementation for Reconfigurable Logic and Interconnects (RIL)-Blocks

The obfuscation with STT-based LUTs, which also uses a sense amplifier, was previously presented. Another STT-MTJ-based LUTs from [9], based on reconfigurable logic and interconnects (RIL)-Blocks that leverage STT-MTJ-based LUTs with MTJ technology and routing-based obfuscation (i.e., switch boxes) will be described. Figure 8.4 illustrates STT-MTJ-based LUT where the WE controls the write operation and $\overline{WE}$ signals, connect each memory cell to the BL and SL, allowing separate access to each memory cell using inputs A and B. The memory content is changed by setting BL and SL during the write operation. Additionally, in each write operation, the content of the MTJs in each memory cell is changed in a complementary fashion, ensuring that MTJ_i and $\overline{MTJ_i}$ always hold opposite values. This approach *eliminates the need for a sense amplifier* with a reference cell, as the complementary values of each cell can be utilized to reliably read the data stored in the main memory cell, made possible by the wide read margin during the read operation enabled by this sensing method.

After the write operation is completed, the data stored in the MTJs can be read by activating the RE and $\overline{RE}$ signals. These signals enable the read path from V^+ to V^-. M-input LUTs utilize 2^M memory cells to implement M-input Boolean functions. A select tree Mux, constructed with pass transistors and transmission gates,[1] is used to select the memory cell holding the correct function value. Below is a 2-input example of our proposed MRAM-based LUT design.

In Fig. 8.4, the WE controls the write operation and $\overline{WE}$ signals, which connect each memory cell to the BL and source line SL. Each memory cell can be accessed separately

[1] Pass transistors are single MOSFET devices used to control signal flow in circuits. Transmission gates, combining NMOS and PMOS transistors, provide better signal integrity by allowing both logic levels to pass without degradation.

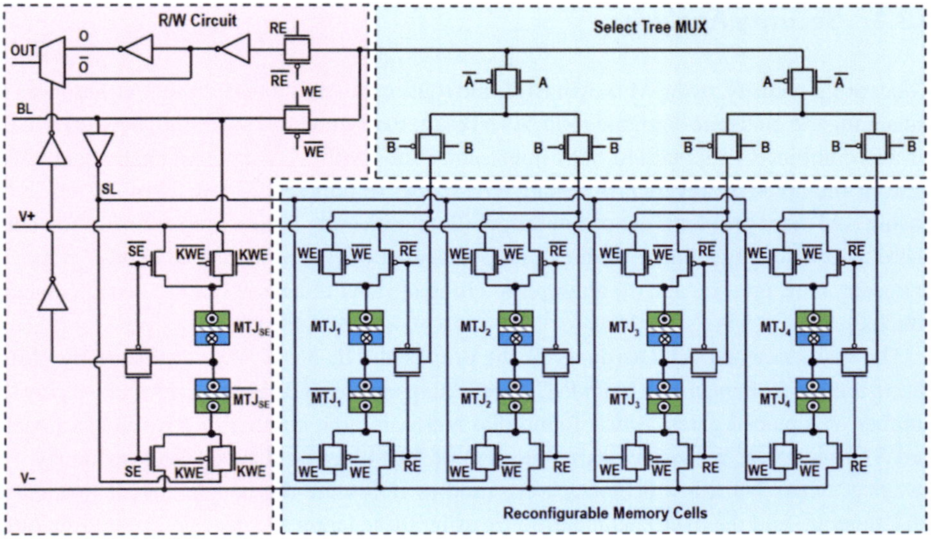

Fig. 8.4 2-input STT-T-based LUT example [9]

and its content can be changed by using the A and B inputs to set BL and SL. During each write operation, the content of MTJs in each memory cell changes in a complementary fashion, ensuring that MTJ_i and MTJ_i always hold opposite values. This eliminates the need for a sense amplifier with a reference cell and allows for reliable data reading due to the wide read margin enabled by this sensing method.

After the write operation is completed and the RE and $\overline{RE}$ signals are asserted, the data stored in the MTJs can be read. The read enable signals RE and $\overline{RE}$ activate the read path from V^+ to V^-, creating a voltage divider circuit. This circuit is used to observe the resistance difference between MTJ_i and MTJ_i. The terminal between the two MTJs is connected to the select tree Mux to direct the appropriate output based on the input signals A and B. The value of the LUT function is observed at the output nodes O and $\overline{O}$.

In order to use a 2×2 RIL-block, a single switch box is needed. The switch box's inputs are connected to the fanout of the LUTs (O_1, O_2) that replace a 2-input gate in the circuit. Additionally, the switch box's output serves as the input to one of the 2-input LUTs. Because the 2×2 RIL-block is vulnerable to instant de-obfuscation using the SAT attack, the size of the switching network is increased to mimic a banyan network. The 8×8 and $8 \times 8 \times 8$ RIL-blocks were implemented. Testing was conducted on both an 8×8 RIL-block with 16 switching elements and an $8 \times 8 \times 8$ RIL-block with 32 switching elements for obfuscation purposes.

8.3.1 Security Analysis

Concerning security, using Muxes in LUT and routing blocks increases the SAT hardness by enhancing the clause-to-variable ratio. As a result, the symmetric switching networks in the circuit, combined with back-to-back interconnections with LUTs, pose a challenge to even state-of-the-art SAT solvers [10]. Using RIL-blocks to build SAT-hard solutions lies in (1) posing SAT-hard problems at each iteration, (2) significantly higher output corruption compared to obfuscating solutions with a one-point function, (3) immunity to bypass, removal, or approximate attacks, and (4) attempts to eliminate SAT attacks and mitigate side-channel attacks.

The resilience and SAT-hardness of the proposed RIL-block were evaluated by obfuscating the *c*7552 benchmark from ISCAS'85 using various RIL-block sizes and varying the number of replaced gates. It was found that increasing the number of RIL-blocks intensified SAT-hardness, while increasing the size of RIL-blocks enhanced SAT-resiliency. For instance, when 3-8 $\times$ 8 $\times$ 8 blocks were used to obfuscate the benchmark, it resulted in SAT-timeout, and the overhead incurred by using these larger blocks was about three times lower compared to using 75-2 $\times$ 2 RIL-blocks. A significant challenge was faced by the SAT solver when larger SAT-hard RIL-block instances were inserted into the circuit – a very deep and large DPLL recursion tree was encountered in each iteration. This resulted in a considerable and exponential increase in the execution time of DPLL recursive calls.

In scan-chain attacks, setting the key in the MTJ to '0' enables the obfuscation circuitry, causing the gate to behave as a NOR gate when the scan enable signal is applied. To prevent scan and shift attacks, key values are stored in the secure cell (SC), which uses a separate scan chain for the SCs. Utilizing complementary MRAM memory cells results in a symmetrical power footprint for read and write operations, maintaining near-zero power variation in the output to mitigate side-channel attacks.

8.4 SOT-MTJ Based LUT Implementation

The utilization of the STT-MTJ-based LUT for obfuscation was previously explained. SOT devices can also be considered as an alternative method for writing data, particularly in *overcoming challenges associated with SST structures*. In the preceding sections, we extensively discussed the topics covered in [3, 11]. Now, we will provide an explanation of SOT devices as reconfigurable logic used for efficient circuit obfuscation. A 2-input SOT-MTJ-based LUT consists of a Mux, spin-hall effect magnetic tunnel junction (SHE-MTJ) write driver, and sensing circuit, as illustrated in Fig. 8.5a. The operations of the SOT-MTJ based LUT involve reconfiguration, which includes programming the SHE-MTJ and storing bits in specific logic states, and computing, which involves executing logic functionality based on input data A and B. During reconfiguration, the write enable (WE) signal is active, while the read enable (RE) signal is inactive. This allows the programming of the SHE-MTJs by

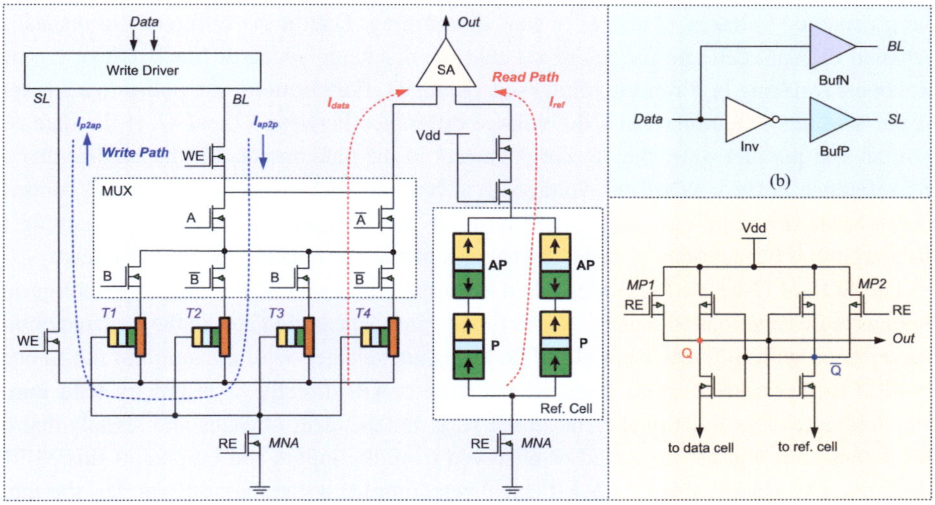

Fig. 8.5 2-input SOT-MTJ based LUT architecture, schematic and circuit diagram of SHE-MTJ write driver [3]

applying the appropriate write current. During the computing procedure, the WE signal is inactive, and the RE signal is active, which leads to the selection of the corresponding MTJ based on the inputs A and B of the LUT. The precharge sense amplifier (PCSA) senses the selected value, generating the appropriate voltage level at the output to represent a logic value.

In the SHE-MTJ write driver, the LUT inputs are utilized to select the MTJ bit cell for programming based on the Mux functionality. The state switching of the MTJ bit cell necessitates a bidirectional current source, which is regulated by a logic unit designed to produce the required current through the selected MTJ bit according to the necessary programming data, as illustrated in Fig. 8.5b. The write path to the BL experiences enhanced driving ability for $P \rightarrow AP$ (slow-write) and $AP \rightarrow P$ (fast-write) by increasing the size of the buffering NMOS transistor in BufN. On the other hand, the driving ability of the write path to the SL is improved by increasing the size of the buffering PMOS transistor in BufP [12]. The SHE effect offers advantages due to its high switching speed (less than 1 ns) and low spin current requirement (less than 60 μA). Furthermore, reconfiguration in serial does not compromise speed, as it only takes a few hundred nanoseconds for the reconfigurations of complex SOT-MTJ-based LUT with more than five inputs.

In Fig. 8.5c, the PCSA consists of two inverters and two PMOS transistors and operates in two phases based on the read enable (*RE*) signal. When *RE* is low, the PCSA precharges the mid-nodes Q and $\bar{Q}$, keeping the amplifier in a metastable state without any stationary current in the circuit. The sensing procedure begins when RE is set to a high voltage, causing the precharged voltages Q and $\bar{Q}$ to start discharging. Due to the difference in resistance between the data cell and the reference cell, the discharge speed differs between the two branches, resulting in different voltages at Q and $\bar{Q}$. Furthermore, the pull-down strength of the inverters is modulated by the voltage difference between Q and $\bar{Q}$. If the data cell is in an anti-parallel state, the discharge current in the data branch will be smaller than in the reference branch. When the voltage Q is charged back to Vdd, and Q will continue to discharge down to Vss. Since there is never any stationary current, only charging or discharging of capacitors, the power consumption can be expected to be nearly zero.

The SOT-MTJ-based LUTs described are used for circuit obfuscation by configuring them as logic functions to replace conventional gates. The logic gates in the targeted circuit are replaced with SOT-MTJ-based LUTs. Gates are selected with a maximum fan-in cone (MFIC) to represent a set of gates whose outputs will directly or indirectly feed into a specified gate. The functionality of an MFIC is independent of gates and signals that do not belong to it and can be tested or observed from the inputs and outputs of this MFIC. Once the gate classification by MFIC has been completed, the method searches the inner gates of each class to identify those suitable for obfuscation, aiming to minimize the design overhead in terms of path delay. The 2-input, 3-input, and 4-input logic gates are replaced with the corresponding LUTs. When obfuscating a gate with more or fewer than two inputs, there is an additional option first to restructure this gate to ensure it ends with a 2-input gate and then obfuscate the final 2-input gate. Additionally, relatively large functional blocks (non-leaf cells) are obfuscated using a single LUT structure, which reduces design overhead compared to obfuscation with multiple combined LUTs.

8.4.1 Results and Security Analysis

The proposed method for obfuscation using SOT-based LUT has been validated on the standard ISCAS'85 and ISCAS'89, MCNC benchmark suites. The open-source ABC program is employed for logic synthesis using TSMC 0.35 μm process design kit (PDK) for overhead measurement.

The approach is resistant to IC testing-based (ITA), circuit partition-based (CPA), brute force (BFA), and side-channel attacks (SCA). SOT-based LUT can withstand restore attacks, making it difficult for attackers to succeed. The complexity of the design is polynomial, which means the obfuscation process can be completed in a reasonable amount of time. Unlike many other obfuscation methods that may have vulnerabilities, this scheme eliminates the possibility of low complexity for attackers. The approach exhaustively enumerates candidate gates for obfuscation to determine if a single input pattern can validate the inputs and sensitize the outputs at the same time. However, it is important to note that this enumeration process has an exponential complexity of 2^m, where m is the number of inputs associated with the gate.

8.5 Transistor-Level Programmable Fabric

Various approaches have been presented that use emerging technologies along with CMOS for design obfuscation. Now, a solution is presented that leverages the *programmability of individual transistors* to obfuscate the design. This purely CMOS-based approach is described in this chapter due to the special features of the programmability at the transistor level. The approach, named "TRAnsistor-level Programmable fabric (TRAP)", consists of logic, interconnect, and memory components. Utilizing a pass transistor-based switch design, routing through one-to-one track links is enabled by TRAP. The design of the TRAP fabric is based on the general principles of the field programmable transistor array (FPTA) as outlined in [13]. Figure 8.6 shows the internal structure of TRAP fabric, including its logic element (LE). At its foundational level, the core entity is the LE, which can be utilized to execute combinational or sequential logic. Comprising four columns of eight transistors (4 vertical + 4 horizontal), the LE can be configured into gates or state-holding components. To enhance performance and area efficiency, each LE also includes built-in FFs, a full adder (FA), and a Mux.

Each LE is integrated with an address decoder, memory blocks, switchboxes, and a repeater to form a unit. The memory blocks store the programming bits, while the switchboxes and repeater transmit these bits to the element. At the highest level, the fabric is structured as an array of functional blocks known as groups. Each group consists of four units, which communicate internally and externally via programmable interconnects with the surrounding circuitry.

The TRAP fabric utilizes a combination of transistors to program logic gates and state-holding components. This integration in the ASIC design flow is made seamless by employing a cell library with a fixed height and variable width. This means that logic gates, which require varying numbers of columns, can begin from any column and extend across logic elements, effectively treating the fabric as a sea of transistors. In the implementation of a NAND3 gate in a logic element, an appropriate bitstream is utilized to activate the required transistors, highlighted in blue, while deactivating the rest. The bitstream also triggers the

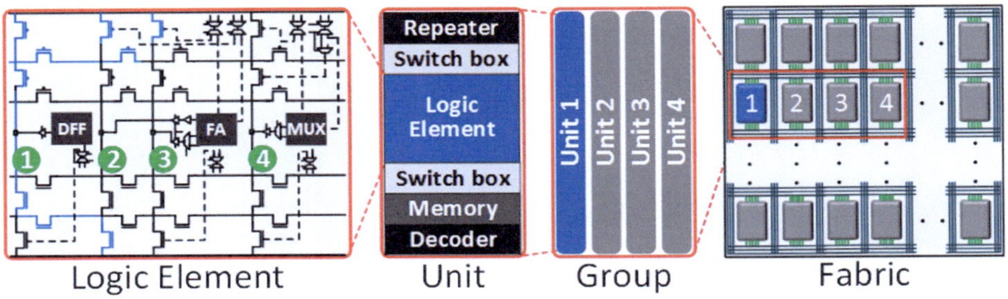

Fig. 8.6 Hierarchical organization of the TRAP fabric [14]

essential multiplexers and switches in the corresponding switchboxes. This approach ensures efficient and reliable operation of the TRAP fabric.

The TRAP fabric supports time-sharing between multiple programs (layers) that can interact and be switched instantaneously to form the desired functionality collectively. This virtualization enables using a smaller TRAP fabric at the cost of additional resources, such as multiplexers, retention FFs, latches, and global signals, to control the layer order. For instance, a four-layer TRAP fabric reduces the number of columns by a factor of 4 compared to a single-layer TRAP. However, the area per column increases by 3.5X due to the additional required resources, resulting in an effective area reduction of 20%. Moreover, programming the four-layer TRAP requires 31 bits per column, while the single-layer TRAP requires only 17, effectively doubling the bitstream size.

This type of ReBO necessitates the use of a custom CAD tool. The ASIC and TRAP components of an obfuscated design are seamlessly integrated by the custom tool for TRAP. In Fig. 8.7, the CAD tool flow can be observed, with the steps exclusively represented by light red ovals and indicated by dark red squares. In the initial stage of the CAD flow, a 65nm TRAP cell library was created, comprising 137 cells. This includes customized cells for DFFs, full adders, and multiplexers. The library was thoroughly characterized and compiled using Synopsys SiliconSmart and Library Compiler. Furthermore, a functionally equivalent cell library for the ASIC part of the design was also developed.

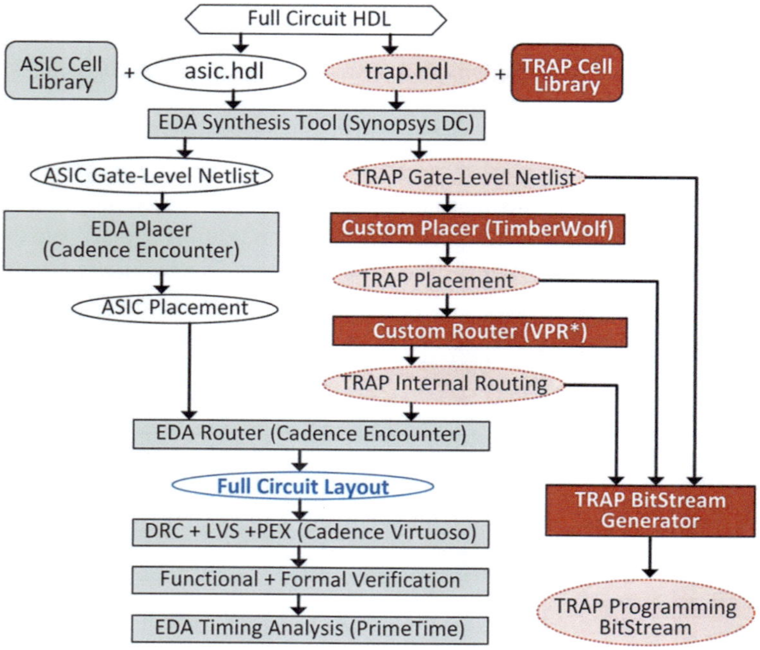

Fig. 8.7 CAD Tool-flow for TRAP-based design obfuscation [14]

First, the design process begins with the creation of the RTL description, encompassing both the TRAP (e.g., 'trap.v') and the ASIC (e.g., 'asic.v') components. Subsequently, the RTL code undergoes synthesis using a specific compiler such as Synopsys Design Compiler. Following synthesis, the ASIC part undergoes placement using a tool like Cadence Innovus, while the TRAP part makes use of the TimberWolf placement and partitioning tool [15]. An initial routing is then carried out for the post-placement TRAP portion to ensure architectural compatibility. This is achieved by utilizing a tailored version of the open-source VPR tool [16]. Concurrently, a custom Python framework processes the netlist, placement, and routing files to generate the programming bitstream for the TRAP portion. Finally, the tool-flow enables timing analysis of the combined ASIC/TRAP design using standard EDA tools.

8.5.1 Results and Security Analysis

The case study of FabScalar RISC-V's architectural pipeline is illustrated to demonstrate the effectiveness of TRAP's obfuscation in terms of PPA and security [17]. The three cases where increasingly more significant portions of the FabScalar processor are obfuscated are presented here. The overhead for implementing architecture map table (AMT), AMT+ result shift register (RSR)+ branch predictor (BP), and Dispatch on TRAP-1L is 1.04X, 1.09X, and 1.20X, respectively. For TRAP, the latency overhead for the AMT, AMT+RSR+BP, and dispatch modules is 1.06X, 1.83X, and 2.64X, respectively. Minimal power consumption overhead is observed with TRAP, with single-layer (TRAP-1L's) power consumption remaining below 1.01X that of a full-ASIC FabScalar implementation in all three cases.

The target modules are obfuscated using both TRAP-1L and TRAP-4L fabrics, and both the brute-force and the SAT-based attack models are considered, assuming that 1 billion combinations can be processed by the attacker per second [14]. In the first scenario, a small yet crucial module of FabScalar's operation is represented by the AMT. Based on the analysis, the AMT, the RSR, and the BP are located across different pipeline stages, with a total design footprint of 8.58%. This leads to the de-obfuscation time for TRAP-1L being increased to 10^{4305} and 10^{1317} hours for brute-force and SAT attacks, respectively. For TRAP-4L, the de-obfuscation times for brute-force and SAT attacks further increase to 10^{7862} and 10^{1319} hours, respectively. Lastly, the entire dispatch stage of FabScalar's pipeline is obfuscated, with a relatively large footprint of 14.34%. In this case, the de-obfuscation time for brute-force and SAT attacks on TRAP-1L is skyrocketed to 10^{9315} and 10^{2871} hours, respectively. TRAP-4L further pushes these times to 10^{16998} and 10^{2873} hours, for brute-force and SAT attacks, respectively.

8.6 Comparative Insights and Discussion

This chapter explored a variety of ReBO techniques leveraging emerging technologies. These techniques generally provide reasonable security. However, this potential for heightened security comes with a set of significant challenges, which we discuss next.

8.7 Comprehensive Security Analysis

The adaptability of LUTs constructed using STT, SOT, and MRAM introduces significant complexity to security assessments. In some cases, it is crucial for potential threats to be able to differentiate between reconfigurable logic and ASIC components. This is similar to other methodologies discussed in Chaps. 6 and 7. Some researchers argue that ReBO engineered with emerging technologies is resistant to SAT attacks and cannot be compromised by them. Additionally, researchers have performed brute force attacks to assess security. Furthermore, most obfuscation techniques lack comprehensive security evaluations against various other forms of attacks, including structural, resynthesis, reverse engineering, as well as side-channel and fault-based attacks. Therefore, there is a need for the development of additional attacks specifically targeting ReBO implemented with emerging technologies, including methods that configure transistors to evaluate their resilience.

8.8 Comparative PPA Analysis

When comparing CMOS technology with MRAM leveraging STT, SOT for PPA overheads, distinct differences emerge, each presenting unique advantages and trade-offs.

In terms of power consumption, CMOS circuits are traditionally known for their low static power due to minimal leakage currents when transistors are in the off state. However, dynamic power consumption in CMOS can become significant, especially in high-frequency operations. As CMOS technology scales down, the increase in leakage currents leads to higher static power consumption, presenting a challenge [18]. On the other hand, STT-MTJ based LUTs and SOT-based LUTs offer lower dynamic power consumption, as they do not require refresh cycles like traditional DRAM. However, STT-MTJ based LUTs require more power during write operations due to the need to switch the magnetic state. SOT-based LUTs, though similar, are more power-efficient in write operations compared to STT-MTJ based LUTs, thanks to their more efficient switching mechanisms. Both STT and SOT MRAM technologies stand out for their extremely low standby power, making them suitable for applications where energy efficiency is paramount.

When it comes to performance, CMOS technology excels in enabling high-speed operations, particularly as it scales down, allowing faster clock speeds in logic circuits. The low latency of CMOS-based SRAM also contributes to its high performance. MRAM-based

LUTs offer relatively fast read and write speeds, but their write speed is generally slower than that of SRAM because of the magnetic state switching process. Despite these advances, CMOS still maintains superior performance for logic operations, while SOT-based LUTs and MRAM-based LUTs show promise for high-performance memory applications with non-volatility.

Regarding area efficiency, CMOS technology remains highly competitive, particularly in logic circuits and SRAM, where transistors can be densely packed, leading to a small footprint. As CMOS technology continues to scale, each transistor's area decreases, although this brings challenges related to variability and leakage. However, the area of CMOS SRAM cells is larger than that of its counterparts, SOT-MTJ and STT-MTJ. This is because the SRAM cell area is more due to the transistor count in comparison to both SOT and STT-MRAM cells. While SOT-MTJ has made strides in reducing cell size compared to STT-MTJ, both still face significant scaling challenges, particularly in maintaining magnetic layer integrity as dimensions shrink. Therefore, while CMOS is more area efficient, particularly for dense logic circuits, MRAM technologies, despite offering non-volatility, currently have a larger area footprint.

8.9 Fabrication Challenges

Fabricating LUTs containing MRAM with STT-MTJ and SOT-MTJ technologies presents unique challenges that differ significantly from those encountered in traditional CMOS processes. The complexity of manufacturing is heightened due to the advanced materials and precise control required for these emerging technologies. Processes such as the deposition of magnetic layers and the formation of tunnel junctions in MRAM are more intricate and less mature than conventional CMOS fabrication techniques. This complexity can lead to increased production costs, lower manufacturing yields, and greater variability in device performance, all of which are detrimental to large-scale commercial deployment.

Another major issue is integration with existing silicon-based technologies. Fabricating MRAM with STT-MTJ and STT-MTJ requires materials and processing techniques that may not be fully compatible with standard CMOS processes. This incompatibility can lead to challenges in achieving reliable and reproducible fabrication outcomes, potentially limiting the scalability and widespread adoption of these technologies [19]. Moreover, the integration process can introduce new sources of defects and variability, further complicating the fabrication and testing of these devices.

References

1. J. Hayakawa, S. Ikeda, Y. Lee, R. Sasaki, F. Matsukura, T. Meguro, H. Takahashi, and H. Ohno, "Current-driven magnetization switching in cofeb/mgo/cofeb magnetic tunnel junctions," in *2006 IEEE International Magnetics Conference (INTERMAG)*, pp. 6–6, 2006.
2. S. Ikeda, K. Miura, H. Yamamoto, K. Mizunuma, H. D. Gan, M. Endo, S. Kanai, J. Hayakawa, F. Matsukura, and H. Ohno, "A perpendicular-anisotropy cofeb–mgo magnetic tunnel junction," *Nature Materials*, vol. 9, pp. 721–724, 2010. Sep
3. J. Yang, X. Wang, Q. Zhou, Z. Wang, H. Li, Y. Chen, and W. Zhao, "Exploiting spin-orbit torque devices as reconfigurable logic for circuit obfuscation," *IEEE Transactions on Computer-Aided Design of Integrated Circuits and Systems*, vol. 38, no. 1, pp. 57–69, 2018.
4. R. Zand, A. Roohi, D. Fan, and R. F. DeMara, "Energy-efficient nonvolatile reconfigurable logic using spin hall effect-based lookup tables," *IEEE Transactions on Nanotechnology*, vol. 16, no. 1, pp. 32–43, 2017.
5. G. Kolhe, T. D. Sheaves, S. M. P. D., H. Mahmoodi, S. Rafatirad, A. Sasan, and H. Homayoun, "Breaking the design and security trade-off of look-up table-based obfuscation," *ACM Trans. Des. Autom. Electron. Syst.*, 2022.
6. T. Winograd, H. Salmani, H. Mahmoodi, K. Gaj, and H. Homayoun, "Hybrid stt-cmos designs for reverse-engineering prevention," in *2016 53nd ACM/EDAC/IEEE Design Automation Conference (DAC)*, pp. 1–6, 2016.
7. *IACR Transactions on Cryptographic Hardware and Embedded Systems*, vol. 2019, p. 97–122, Nov. 2018.
8. T. Winograd, H. Salmani, H. Mahmoodi, K. Gaj, and H. Homayoun, "Hybrid STT-CMOS designs for reverse-engineering prevention," in *Proceedings of the 53rd Annual Design Automation Conference*, pp. 1–6, 2016.
9. G. Kolhe, S. Salehi, T. D. Sheaves, H. Homayoun, S. Rafatirad, M. P. Sai, and A. Sasan, "Securing hardware via dynamic obfuscation utilizing reconfigurable interconnect and logic blocks," in *2021 58th ACM/IEEE Design Automation Conference (DAC)*, pp. 229–234, IEEE, 2021.
10. A. Biere, T. Faller, K. Fazekas, M. Fleury, N. Froleyks, and F. Pollitt, "CaDiCaL 2.0," in *Computer Aided Verification - 36th International Conference, CAV 2024, Montreal, QC, Canada, July 24-27, 2024, Proceedings, Part I* (A. Gurfinkel and V. Ganesh, eds.), vol. 14681 of *Lecture Notes in Computer Science*, pp. 133–152, Springer, 2024.
11. A. Brataas and K. M. D. Hals, "Spin–orbit torques in action," *Nature Nanotechnology*, vol. 9, pp. 86–88, 2014. Feb
12. R. Bishnoi, M. Ebrahimi, F. Oboril, and M. B. Tahoori, "Improving write performance for stt-mram," *IEEE Transactions on Magnetics*, vol. 52, no. 8, pp. 1–11, 2016.
13. J. Tian, G. R. Reddy, J. Wang, W. Swartz, Y. Makris, and C. Sechen, "A field programmable transistor array featuring single-cycle partial/full dynamic reconfiguration," in *Design, Automation & Test in Europe Conference & Exhibition (DATE), 2017*, pp. 1336–1341, 2017.
14. M. M. Shihab, B. Ramanidharan, S. S. Tellakula, G. Rajavendra Reddy, J. Tian, C. Sechen, and Y. Makris, "ATTEST: Application-agnostic testing of a novel transistor-level programmable fabric," in *2020 IEEE 38th VLSI Test Symposium (VTS)*, pp. 1–6, 2020.
15. T. W. S. Inc., "Timber wolf eda tool." http://twolf.com/, 2024. Accessed: July 28, 2024.
16. K. E. Murray, O. Petelin, S. Zhong, J. M. Wang, M. Eldafrawy, J.-P. Legault, E. Sha, A. G. Graham, J. Wu, M. J. P. Walker, H. Zeng, P. Patros, J. Luu, K. B. Kent, and V. Betz, "VTR 8: High-performance cad and customizable FPGA architecture modelling," *ACM Transactions on Reconfigurable Technology and Systems*, vol. 13, no. 2, 2020.

17. N. K. Choudhary, S. V. Wadhavkar, T. A. Shah, H. Mayukh, J. Gandhi, B. H. Dwiel, S. Navada, H. H. Najaf-abadi, and E. Rotenberg, "Fabscalar: Composing synthesizable rtl designs of arbitrary cores within a canonical superscalar template," in *2011 38th Annual International Symposium on Computer Architecture (ISCA)*, pp. 11–22, 2011.
18. R. Saha, Y. P. Pundir, and P. Kumar Pal, "Comparative analysis of stt and sot based mrams for last level caches," *Journal of Magnetism and Magnetic Materials*, vol. 551, p. 169161, 2022.
19. Z. U. Abideen, S. Gokulanathan, M. J. Aljafar, and S. Pagliarini, "An overview of FPGA-inspired obfuscation techniques," *Association for Computing Machinery*, vol. 56, no. 12, December 2024.

Part II
Balancing PPA and Security: A Hybrid ASIC Approach

Part II of this book focuses on an approach termed hybrid ASIC (hASIC) that introduces a unique solution using a combination of ASIC and reconfigurable logic. In general, only the most sensitive circuit part is placed on the reconfigurable part, minimizing the PPA overheads. The part that remains in the ASIC has to designed such that it does not leak sensitive information or traces for adversaries. hASIC uses fine-grain obfuscation leveraging LUTs. However, implementing this approach involves integrating a custom tool into the traditional CAD flow, which can be challenging. To address this challenge, a framework is necessary to enable designers to make informed decisions and manage trade-offs effectively. Such a platform would help in evaluating design versus security trade-offs and assessing the impact of implementing security measures.

To achieve this, we developed a custom tool named *Tunable Design Obfuscation Technique (TOTe)* to obfuscate a design with a given obfuscation rate. The summary of the chapters is given as follows:

Chapter 9 introduces the design obfuscation concept, focusing on the trade-offs between design and security considerations. It discusses the security-aware CAD flow, including stages such as RTL code generation, logic synthesis, and physical synthesis. It also presents the TOTe tool and its initial results, focusing on PPA overheads. Moreover, it introduces techniques for decomposing LUTs and enhancing QoR, as well as the analysis and experimental findings using TOTe.

Chapter 10 demonstrates the physical implementation of hybrid ASIC using commercial CMOS technology. It presents the physical implementation of selected designs, including large circuits with different levels of obfuscation. Additionally, it showcases the final layouts for baseline and optimized variants, emphasizing the improvements achieved through optimization techniques presented in the previous chapter.

Chapter 11 provides a thorough examination of the threat model and security implications of hASIC-obfuscated circuits. It explores different types of attacks, such as oracle-guided and oracle-less attacks, to gauge the security of the obfuscated designs. Additionally, it offers a comprehensive overview of design obfuscation and custom structural attacks to assess the efficacy and resilience of obfuscated circuits.

A Security-Aware CAD Flow for Hybrid ASIC

9

9.1 Design Obfuscation Concept

The section discussed the hASIC design obfuscation concept that is a form of reconfigurable-based obfuscation. Figure 9.1a shows an ASIC, a static design that offers no reconfiguration but best-in-class performance. Let us focus on FPGA as highlighted in Fig. 9.1d. The fabric of an FPGA device typically includes multiple reconfigurable blocks. FPGA is a flexible device that can be considered a fully obfuscated hardware of sorts. An FPGA incurs performance penalties with respect to an ASIC. If the fabric from Fig. 9.1d is taken and embedded into the design represented in Fig. 9.1a, a new device is formed, as shown in Fig. 9.1c. This device is still an ASIC but now containing an embedded FPGA and therefore implementing eFPGA-based obfuscation. However, even embedding relative small eFPGA macros leads to PPA overheads compared to ASIC.

Most techniques aim to minimize the reconfigurable part to avoid significant performance and area overheads. Unfortunately, this compromises the security of the obfuscated design. To adjust the level of obfuscation, the reconfigurable tile in Fig. 9.1c could be increased. However, distributing the logic all over the fabric or utilizing fine-grain reconfigurable logic, as shown in Fig. 9.1b, where LUTs and standard cells are entirely mixed, can improve PPA and offer more flexibility. This book takes advantage of this possibility along with the reconfigurable elements. Moving from right to left, performance increases, while obfuscation and flexibility increase from left to right. However, neither extreme is an ideal design point for circuits with strict security and performance requirements. A middle solution that can balance performance and security is **hASIC**.

Figure 9.1b illustrates a hybrid design incorporating fine-grain reconfigurable and static parts to achieve obfuscation. The reconfigurable part offers security in the circuit, while the static logic provides performance benefits. The generated block architecture consists of a combination of reconfigurable and static cells to explore the design space. Programmable

Z. U. Abideen and S. Pagliarini, *Reconfigurable Obfuscation Techniques for the IC Supply Chain*, Synthesis Lectures on Digital Circuits & Systems,
https://doi.org/10.1007/978-3-031-77509-3_9

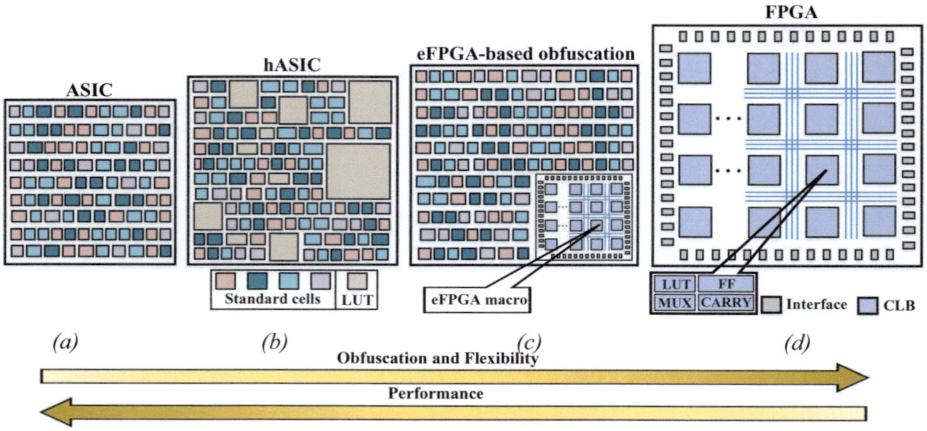

Fig. 9.1 The design obfuscation landscape

LUTs are utilized to implement the reconfigurable part, rendering the circuit non-functional until programmed. However, this book also emphasizes the importance of a *high degree of obfuscation* in effectively securing the circuit, generally not investigated in state-of-the-art techniques.

9.2 Security-Aware CAD Flow for hASIC

The security-aware CAD flow is depicted in Fig. 9.2, which utilizes a custom tool to generate an hASIC design with static and reconfigurable logic. The entire obfuscation process is automated; Therefore, the design time experiences a slight increase when compared to the traditional ASIC flow. During the initial phase of the process, a commercial synthesis tool for FPGA creates a Verilog netlist of the targeted design. Then, the FPGA synthesis tool generates a timing report and a netlist that includes standard FPGA primitives like LUTs, MUXes, and FFs. To achieve its primary goal of replacing FPGA cells with ASIC cells, TOTe utilizes the ASIC standard cell library of choice, the outputs generated by FPGA synthesis and a user-defined obfuscation target obf_c.

In the second phase, a parser reads elements from the netlist and paths from the timing report. These are then sent to a timing engine for processing. Then, it selects LUTs in the critical path and replaces them with standard cells that implement the same logic, removing the programmability aspect. In other words, TOTe recognizes critical paths and replaces 'slow' reconfigurable elements with 'fast' static ones. This process is repeated until no more LUTs can be converted in order to respect the user-defined obfuscation target obf_c. The obfuscation target controls the ratio of LUTs that must remain programmable. By replacing LUTs with static logic, TOTe reduces the area, power, and delay, thereby improving the

Fig. 9.2 A security-aware CAD flow for hASIC

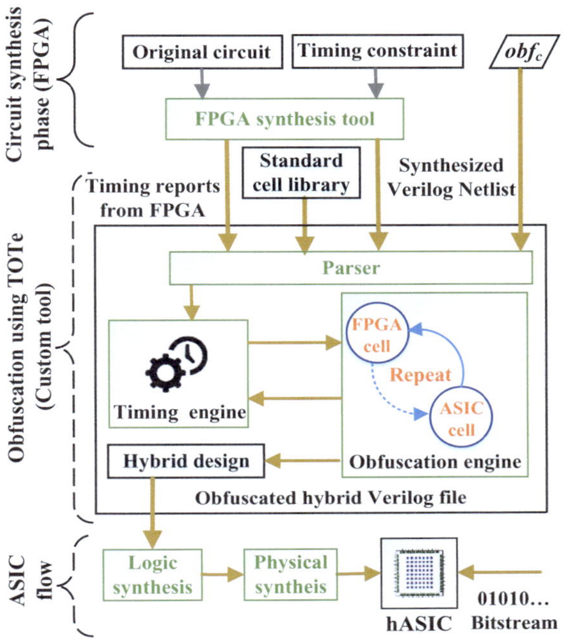

frequency of the design. The output of the tool is an obfuscated *hybrid* Verilog file containing both reconfigurable LUTs and a static part. Finally, to complete the hASIC design, physical synthesis is performed to generate the layout. The resulting layout is then sent to the foundry for fabrication. The following subsection provides a comprehensive overview of the flow and internal architecture.

9.2.1 Detailed Flow and Internal Architecture of ToTe

The process of obfuscating a design and producing an hASIC through logical and physical synthesis involves a comprehensive 7-step approach, as depicted in Fig. 9.3. Circled numbers represent these steps in the following text.

During Step ① of the process, the original circuit is synthesized using a commercial FPGA synthesis tool. Notably, the original circuit requires no special annotations, synthesis pragmas, or any other changes in its representation. The output of this step consists of a synthesized netlist and a timing report, with the netlist comprising all the typical FPGA primitives such as MUXes, LUTs, and FFs. It is important to note that at this point, the logic of the design is 100% obfuscated since the design entirely consists of LUTs.

Moving on to Step ②, the pre-processing stage begins with filtering and interpreting the timing report and Verilog netlist. Parsing the timing report is relatively straightforward

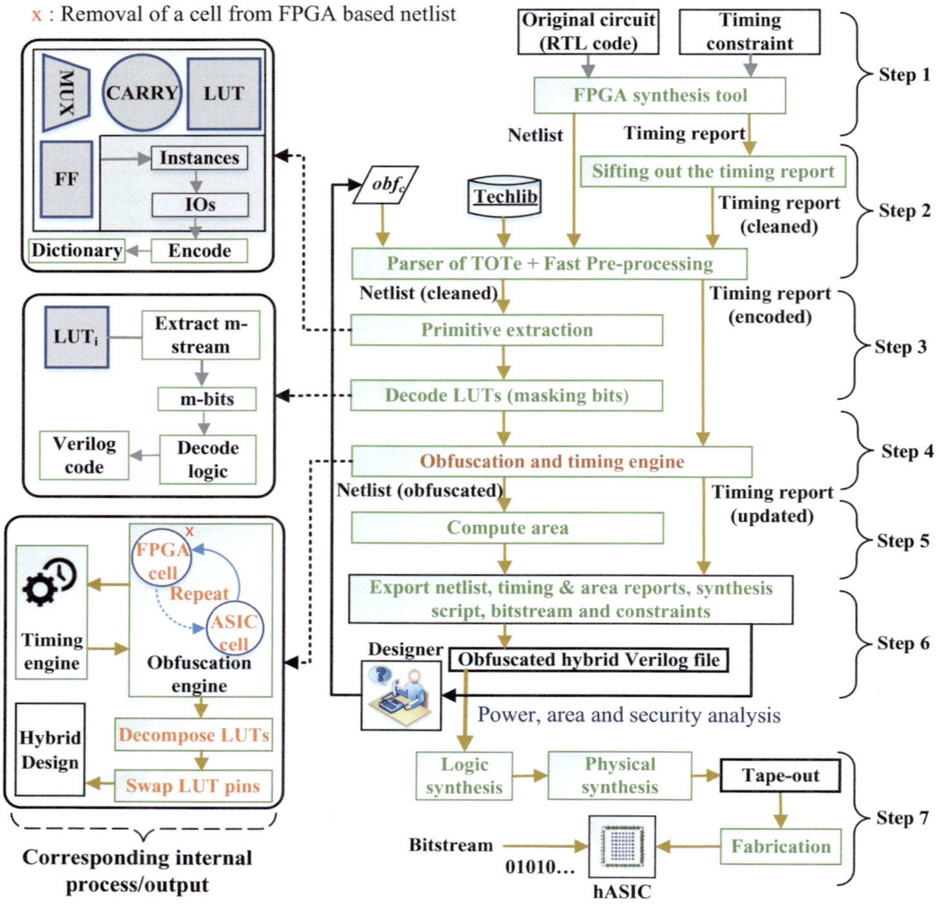

Fig. 9.3 Overview of the obfuscation flow and its inner steps

once some redundant information is discarded. Once the timing report has been filtered, each analyzed path may contain four FPGA primitives, namely FF, $CARRY$, LUT_i, and MUX. TOTe encodes (hashes) instance names to ensure a more efficient representation, reducing the need for lengthy string representations. The pre-processing stage concludes when it generates a list of timing paths, where each path consists of a collection of hashed instances and their corresponding delay values. It is important to note that an instance may appear in several paths and under different timing arcs. Finally, the list of timing paths is sorted in ascending order. The path with the highest delay, the Critical Path (CP), is used as a reference. The sum of all CPs is referred to as $sumCP$.[1]

In Step ③, TOTe performs primitive extraction and LUT decoding to preserve the circuit structure after optimization. To achieve this, the tool creates a graph representation of the netlist to keep track of port connections. Every instance is annotated with its primitive type, including configuration bits for LUTs. TOTe can interpret the LUT encoding scheme used in the FPGA netlist. For example, when dealing with a LUT_6, the tool extracts a 64-bit masking pattern from the netlist, which is then converted into a truth table with six inputs and one output. Figure 4.2 shows the truth table for LUT_2. The masking pattern determines which input combinations generate 1s and 0s at the output. Using the populated truth tables, the tool builds combinational logic equivalent to the LUT's logic. Finally, the truth tables are exported as synthesizable Verilog code. For other primitives, such as FF and MUX, no decoding is required, and they are directly translated into their ASIC equivalent logic cells.

The security and performance objectives of the tool are driven by obfuscation and timing engines. These engines are responsible for various important tasks, such as critical path identification, timing analysis, and replacement of reconfigurable cells for static cells, and are utilized in Step ④. Algorithm 9.1 outlines the various operations performed within the obfuscation algorithm of the tool. In this algorithm, the list of LUTs is denoted as L, the list of timing paths as P, and the obfuscation level as obf_c. Additionally, L_{ST} and L_{RE} are internal variables that represent lists of static and reconfigurable LUTs, respectively. At the start of the obfuscation algorithm, all LUTs are stored in L_{RE} on line 1. It calculates the value of the K variable in terms of the number of LUTs to be realized as a static part on line 2, where the SIZE_OF function returns the number of elements in the list. In the loop that appears on lines 3–9, the critical path is identified on line 4 using the FIND_CRITICAL function, and the slowest LUT on that path is identified using the FIND_SLOWEST function on line 5. If the identified LUT is in L_{RE} on line 6, the lists of LUTs are updated on lines 7–8, where the INSERT and REMOVE functions insert and remove the LUT, respectively. The timing engine recalculates the affected paths on line 9, where the UPDATE_INSTANCENAME_TIMING function updates the instance name of the corresponding LUT as a static logic and the critical path and delay of the corresponding LUT in the timing report. This loop on lines 3–9 continues until K LUTs are selected for the static part.

Additional steps on lines 10–17 are required to implement hASIC. The DECODE function on line 11 operates on each LUT that was mapped as a static part. The description of these LUTs in Verilog as truth tables is already processed during Step ③. Subsequently, the ASIC synthesis of the truth tables is executed to obtain netlists composed of standard cells. Timing and power analysis during physical synthesis is generated by the function GEN_CASE_0_1

[1] It is worth noting that CP and $sumCP$ are analogous to Worst Negative Slack (WNS) and Total negative Slack (TNS) in traditional Static Timing Analysis (STA), except that all paths in this analysis are assumed to pass timing checks, which means that no negative values are considered for the sake of simplicity.

Algorithm 9.1: Obfuscation procedure

Input: L, P, obf_C
Output: $hASIC$
1 $L_{ST} \leftarrow \phi$, $L_{RE} \leftarrow L$
2 $K \leftarrow (1 - obf_c) \times SIZE\_OF(L_{RE})$
3 **while** $SIZE\_OF\ (L_{ST}) \leq K$ **do**
4     $path \leftarrow$ FIND_CRITICAL(P)
5     $lut \leftarrow$ FIND_SLOWEST($path$)
6   **if** $lut \in L_{RE}$ **then**
7         $INSERT(lut, L_{ST})$
8   REMOVE(lut, L_{RE})
9   UPDATE_INSTANCENAME_TIMING(lut, P)
10 **for** *each* $lut \in L_{ST}$ **do**
11     $Design_{ST} \leftarrow$ DECODE(lut)
12 **for** *each* $lut \in L_{RE}$ **do**
13   GEN_CASE_0_1(lut)
14   DECOMPOSE_OPT(lut)
15   SWAP_PINS(lut)
16 Design_RE ← GEN_RE(L_RE)
17 **return** $hASIC \leftarrow Design_{ST} \cup Design_{RE}$

for ‘force logic’ on line 13. Without this setting, each LUT would be timed for its worst timing arc instead of the implemented timing arc when the LUT is programmed with the target design.

Additionally, larger LUTs are decomposed into smaller LUTs by DECOMPOSE_OPT on line 14, which will be described in Sect. 9.3. On line 15, SWAP_PINS performs a final timing optimization that attempts to swap the LUT pins to improve the delay, which is also discussed later in Sect. 9.3.4. The function GEN_RE on line 16 generates the reconfigurable part. Ultimately, the algorithm merges the design generated for the static part, $Design_{ST}$, and the reconfigurable part, $Design_{RE}$, to build hASIC.

During Step (5) of the obfuscation process, the tool estimates the area of the hASIC design using the formula $A = A_{RE} + A_{ST}$. To determine the area of the reconfigurable part, denoted as A_{RE}, it sums up the area of the reconfigurable LUTs. Similarly, it computes the area of the static part, denoted as A_{ST}, by summing up the area of the standard cells of all the static LUTs. It uses an industry-grade physical synthesis tool that properly considers congestion to ensure a highly accurate estimate. In the hASIC design process, Step (6) involves creating files that describe hASIC. This step generates an obfuscated hybrid Verilog file, timing, and area reports. Designers can repeat this process until they achieve their desired level of obfuscation and performance.

Finally, in Step (7), the obfuscated netlist is synthesized using a commercial synthesis tool. hASIC is implemented using a commercial physical synthesis tool, which executes traditional Place & Route (P&R), CTS (Clock Tree Synthesis), Design Rule Check (DRC), and other necessary steps. The resulting tapeout database is then sent to the foundry for fabrication. Once the fabricated parts are received, programming is required for the hASIC design to function correctly, which involves using a bitstream, just like in an FPGA design.

9.3 Building Custom LUTs

This section discusses optimizations related to LUTs and the measures taken to ensure that an hASIC design exhibits an ASIC's high-performance attributes while maintaining the obfuscation capabilities of an FPGA fabric.

9.3.1 Standard Cell Based LUTs

LUTs of different input sizes (LUT_1, LUT_2,..., LUT_6) have been designed from *regular standard cells*, following the Versatile Place and Route's (VPR) template [1]. Table 9.1 shows the area, density, number of FFs, combinational cells, and average delays of the implemented LUTs. The area for the macros that represent the LUTs approximately doubles from LUT_i to LUT_{i+1}. The number of flip-flops grows with the LUT_i size (2^i). The average delay highlights the average of all timing arcs. It should be emphasized once more that the LUTs were generated as macros composed of standard cells, thereby rendering them compatible with traditional standard cell based design flows. The LUTs are highly compact, with the main design goal being area/density. The layouts for LUT_4, LUT_5, and LUT_6 macros are shown in Fig. 9.4.

Commercial FPGAs typically have limited flexibility in terms of implementing LUTs of different sizes. However, hASIC can implement designs with different LUT sizes due to its design-specific nature. This means that the reconfigurability aspect of FPGAs is no longer entirely necessary. We are not interested in building a fabric that works well for the average design. Instead, we seek a combination of LUT sizes that works well for the design at hand.

Table 9.1 Block implementation results for LUTs

Macro	Area (μm^2)	Density (%)	# FFs	Comb. cells	Avg. delay (ns)
LUT_1	36.00	76.00	2	1	0.049
LUT_2	64.80	76.26	4	1	0.052
LUT_3	117.00	89.23	8	8	0.119
LUT_4	259.20	85.23	16	15	0.192
LUT_5	491.40	91.50	32	33	0.257
LUT_6	957.60	91.09	64	36	0.295

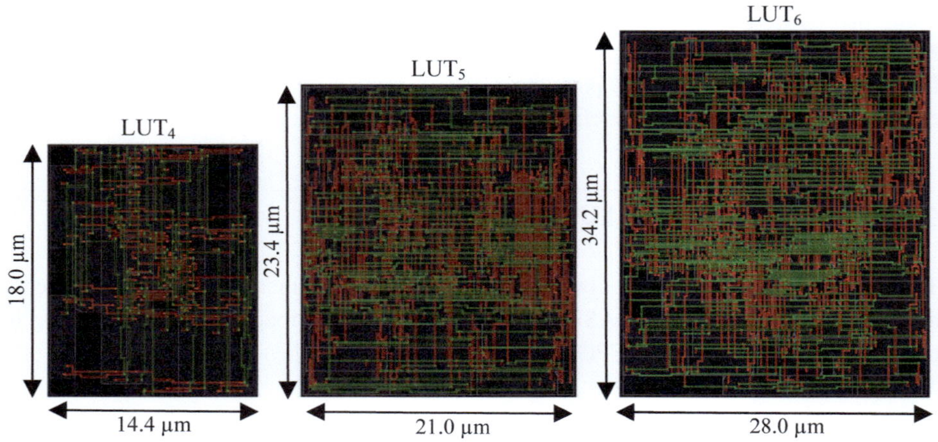

Fig. 9.4 The layout of macros for LUT_4, LUT_5, and LUT_6 [2]

Additionally, LUT macros are highly compact, allowing for high-density designs. Each LUT includes FFs for storing configuration bits that serve as a lock for the obfuscated design and three extra pins for configuring the registers (*serial_in*, *serial_out*, and *enable*). The LUTs are connected in a serial chain, similar to a scan chain. Using FFs, the technology-agnostic framework makes floorplanning and placement effortless. Furthermore, the LUT macros are treated as regular standard cells during physical synthesis, allowing TOTe to take full advantage of commercial EDA tool placement algorithms and eliminating the need for custom scripts for placing the LUT macros.

9.3.2 LUT Decomposition

The area and delay of a LUT are directly correlated with its number of inputs. The size is primarily determined by the number of sequential elements needed to store the LUT's truth table, while the speed is proportional to the LUT's internal MUX tree. However, not all 6-input functions need a LUT_6 for implementation. For example, an AND6 can be broken into 5 AND2s, as shown in Fig. 9.5. According to Table 9.1, it is evident that the area almost doubles for each additional input. The delay increases significantly, with a LUT_6 having almost six times the average delay of a LUT_2. The example demonstrates that decomposing a LUT_6 can reduce the area to less than one-third of its original size. Furthermore, the delay is reduced to approximately half. This example illustrates how decomposition is a promising approach for *improving delay and reducing area*. It will also reduce the power observed during the physical synthesis. To decompose LUTs, TOTe utilizes Functional Composition (FC) [3]. This approach enables bottom-up association of Boolean functions and offers control over the costs involved in the composition process.

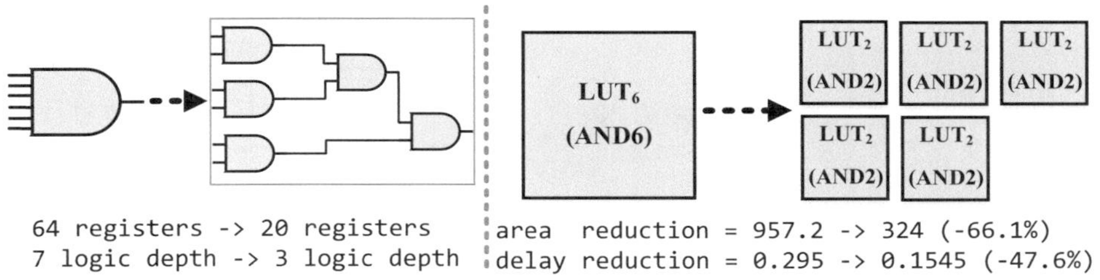

Fig. 9.5 Logic conversion and decomposition of LUT_6 [2]

9.3.3 Functional Composition for LUTs

This section will provide an overview of the functional composition (FC) paradigm and how it can be applied to LUTs. Readers can refer to the sources listed in [3, 4] for more in-depth information. The FC paradigm is a bottom-up approach guided by five core principles. First, it uses bonded pairs (BPs) that consist of a functional part (a canonical implementation of a Boolean function, such as a binary decision diagram or truth table) and an implementation part (the structure being optimized, such as a fanout-free LUT circuit). First, each BP association performs independent functional/implementation operations, which allows for more complex implementations with simpler functional operations. Second, using partial ordering, all BPs with the exact cost are stored together in a set (bucket), enabling intermediate solutions as sub-problems and associations' performance in a cost-increasing fashion. Third, initial BPs, such as constants and single input variables, are required to initiate any FC algorithm. Finally, the FC paradigm allows the heuristic selection of a subset of permitted functions to reduce the composition search space.

9.3.4 Pin Swap Approach

The LUTs mentioned in Sect. 9.3.1 are essentially a MUX tree fed by registers storing a truth table that can be customized. The MUX tree is the primary factor that affects the LUT delay, especially for LUTs with multiple inputs. Therefore, the order in which the pins are arranged in the MUX tree significantly impacts the LUT delay. Inputs connected to a MUX closer to the output will have lower logic depth and smaller delays.

The SWAP_PINS method used in Algorithm 9.1 utilizes the flexibility of LUT functions to allow for arbitrary input pin swaps by permuting the function's truth table. This approach uses a LUT function and timing information and outputs the permuted truth table and a new order of input pins/nets. The example in Fig. 9.6 demonstrates a successful pin swap that improves the timing slack of a design. The pin swap algorithm considers the LUT function, Arrival Time (AT) for each input net (referred to as $[n0, n1, n2]$), the cell arc delay (DLY) associated with each input, and the required time (RT) at the output. In the presented

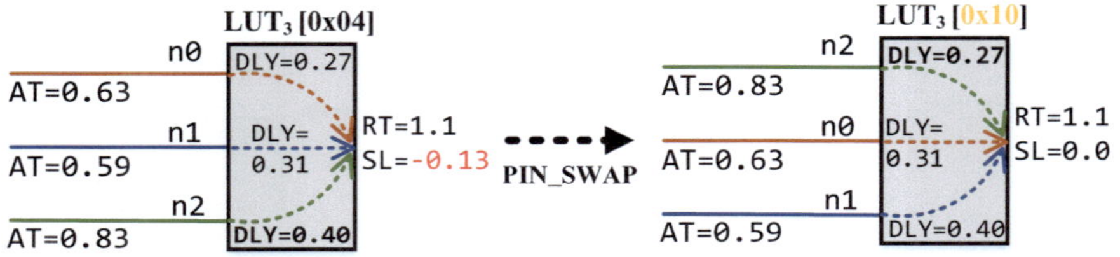

Fig. 9.6 Example of a beneficial pin swap [2]

example, the critical arc is $n2$, with a total delay of 1.23, and RT $= 1.1$. The algorithm explores all input permutations to minimize WNS, and if negative slack is detected in two or more arcs, it also attempts to reduce TNS. If a new order improves WNS and/or TNS, the truth table is permuted to maintain the same functionality. The algorithm outputs the truth table $0x10$ and the new net order $[n2, n0, n1]$.

9.4 Experimental Results

This section presents a comparison between performance and security trade-offs for various designs at different degrees of obfuscation. The objective is to present a range of representative designs, including established benchmarks and circuits for different applications such as crypto cores, filters, CPU, GPU, etc. The designs were selected for their diverse architecture, applications, and number of LUTs in the critical path. All experimental results were obtained by executing FPGA synthesis in Vivado, with the target device set to Kintex-7 XC7K325T-2FFG900C containing 6-input LUTs. Subsequently, Cadence Genus was employed for logic synthesis, using three flavors of a commercial 65nm standard cell library (LVT/SVT/HVT). It is important to note that TOTe is completely *agnostic with respect to PDKs, libraries, and tools* and that these results can be generalized to other technologies and CAD vendors.

For the initial experiment, it was desired to cover all possible FPGA primitives with a compact design. A schoolbook multiplier (SBM) design [6] was utilized for this purpose. To analyze the effects of obfuscation on performance and area trends, an 8-bit SBM has been obfuscated by varying obf_c from 55 to 100%. The synthesis targeted a challenging frequency of 540MHz, and the timing engine calculations showed that CP and $sumCP$ values became 0.490 ns and 16088.69 ns, respectively. The values are obtained for a design at 100% obfuscation level, representing a design analogous to an FPGA. It is important to note that TOTe implements a simple timing analysis. Yet, during the obfuscation process, CP and $sumCP$ remained consistent among themselves. This consistency is sufficient to determine critical paths generally, and realistic timing values can later be obtained during the final timing analysis using a commercial physical synthesis tool.

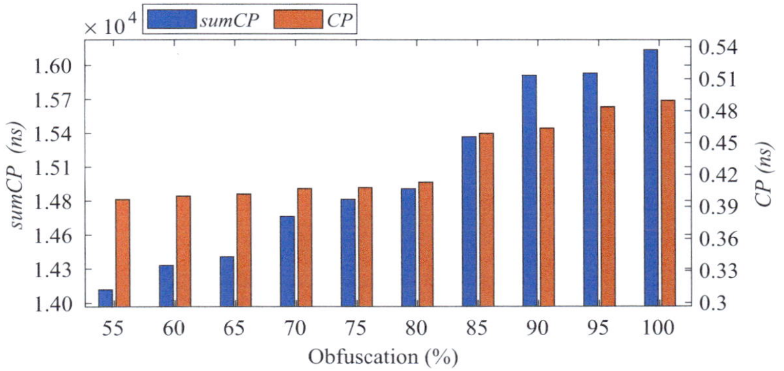

Fig. 9.7 Obfuscation versus performance trade-off for SBM [5]

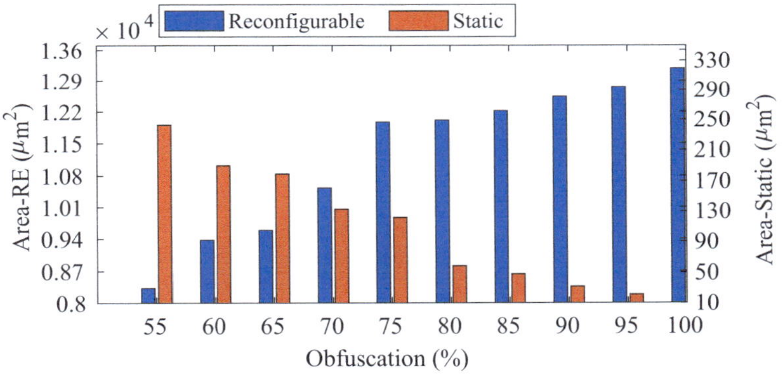

Fig. 9.8 Obfuscation versus area trade-off for SBM [5]

The performance of the 8-bit SBM was analyzed after performing obfuscation at different levels. The timing characteristics are illustrated in Fig. 9.7. The results showed that increasing the level of obfuscation led to a decrease in performance and vice versa. The trend depicted in Fig. 9.7 was that CP improved inversely with the obfuscation, but it saturated when the obfuscation was below 80%. However, this was not the case for $sumCP$, as the decrease in obfuscation caused continuous improvement. The obfuscation versus area profile of the 8-bit SBM is also illustrated in Fig. 9.8.

An investigation was carried out to determine if similar saturation would be exhibited by other designs. To accomplish this, the ISCAS'85 benchmarks were selected and the results are presented in Fig. 9.9. These combinational benchmarks were chosen as they have only one stage of logic, making it easier to trace the correlation between CP and $sumCP$ (i.e., the critical path remains unchanged irrespective of different reg2reg paths). Surprisingly, even in these simple designs, saturation occurs remarkably fast. Obfuscation has also been applied to more representative designs to cover more comprehensive results. The results for

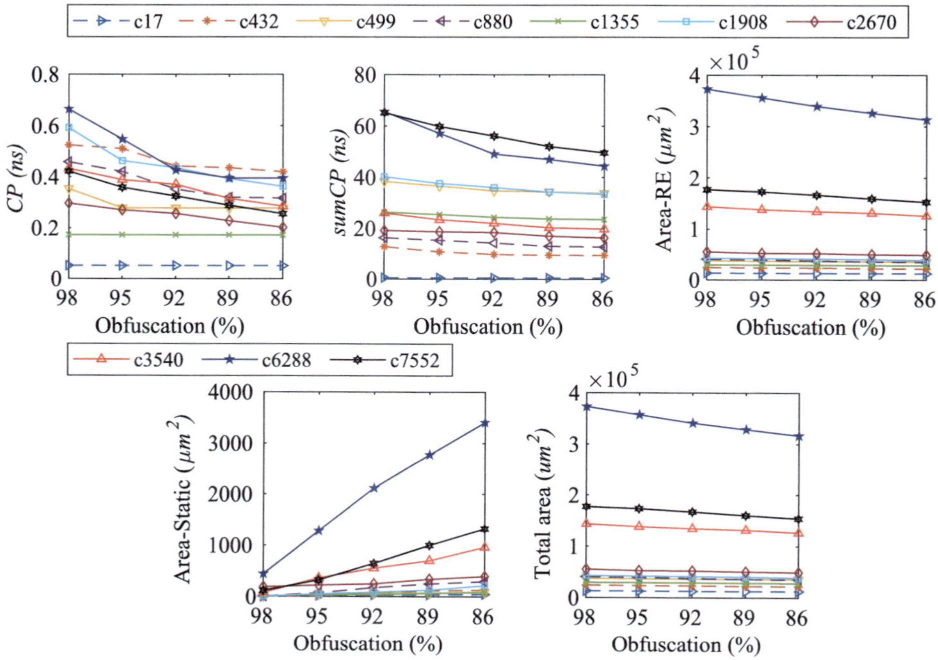

Fig. 9.9 Obfuscation results for ISCAS'85 benchmarks [2]

IIR, PID, Median Filter, SHA-256, and other cryptocores and large designs are presented in detail in Table 9.2. The designs listed in Table 9.2 are sorted based on the number of LUTs used. Graphical representations of the results for AES, RISC-V, and SHAKE-256 designs have also been included in Fig. 9.10, which provide a visualization of the trends. Regarding optimization, the results for the decomposed LUTs will be presented later, where physical synthesis will be executed for a fair comparison between the baseline and optimized designs.

To summarize, the findings presented in this chapter validate the trade-offs between design and security for numerous designs. It is evident that using a LUT-based circuit representation, similar to an FPGA, affects delay and area differently. Regarding area, the trend is straightforward - the smaller the obfuscation target, the more compact the circuit. However, when it comes to delay, it seems that hASIC incurs performance penalties that reducing the targeted obfuscation level alone cannot overcome. Therefore, the functional decomposition of LUTs is employed to achieve better performance. The next chapter will present a detailed physical synthesis analysis, including applying optimization methods described in Sect. 9.3 to enhance performance.

Table 9.2 Detailed results for selected designs

Design	Obf. (%)	*sumCP* (*ns*)	*CP* (*ns*)	Area-RE (μm^2)	Area-ST (μm^2)	LUT (RE)	LUT (ST)
IIR [7]	98	1574.48	0.591	55031.04	257.4	584	11
	95	1553.32	0.526	54104.40	720.72	566	29
	92	1534.39	0.526	53177.76	1184.04	548	47
	89	1501.29	0.526	52251.36	1647.36	530	65
	86	1489.93	0.526	51324.48	2110.68	512	83
PID [8]	98	2547.58	0.756	445590.00	2816.82	896	18
	95	2466.25	0.642	432340.92	9441.36	869	45
	92	2391.96	0.592	421365.95	14928.84	841	73
	89	2348.61	0.568	407273.76	21974.94	814	100
	86	2322.46	0.543	392345.64	29439.00	787	127
Median Filter [9]	98	637.584	0.963	499601.16	2860.2	979	19
	95	563.584	0.747	483561.00	10880.28	949	49
	92	504.805	0.597	469323.72	17998.92	919	79
	89	480.018	0.543	448850.52	28235.52	889	109
	86	466.454	0.543	427346.27	38987.64	859	139
SHA-256 [10]	98	7425.73	0.962	1313150.76	10291.86	2195	44
	95	7354.59	0.871	1275984.00	28875.24	2128	111
	92	7322.15	0.871	1233448.56	50142.96	2060	179
	89	7301.94	0.871	1179674.64	77029.92	1992	246
	86	7164.02	0.871	1125799.56	103967.46	1925	313
FPU [11]	98	2909.06	0.707	1031676.84	1250.028	2487	50
	95	2734.00	0.650	1003225.68	2672.586	2412	126
	92	2572.95	0.650	966715.20	4498.11	2336	202
	89	2478.73	0.650	935060.04	6080.868	2259	279
	86	2410.21	0.650	893005.56	8183.592	2183	355
GPU [12]	98	237699.30	0.933	21317740.2	124971.48	40739	831
	95	215696.68	0.871	21009821.4	278931.10	40102	2078
	92	185520.65	0.750	20015822.2	495521.11	39492	3352
	89	154560.56	0.650	19552256.6	781521.30	38243	4521
	86	135802.32	0.625	18552023.3	1011230.2	36125	5806

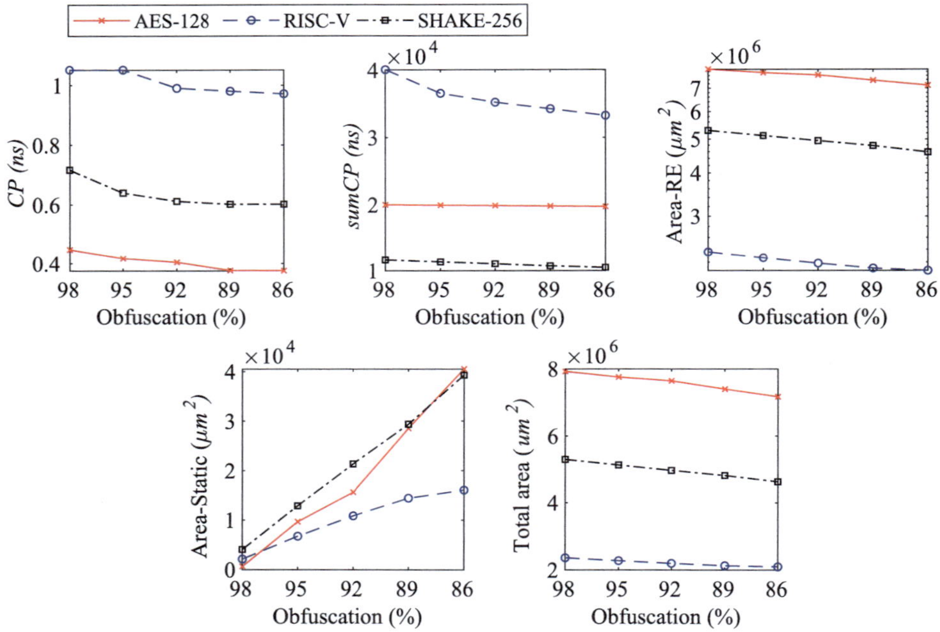

Fig. 9.10 Obfuscation results for AES-128, RISC-V and SHAKE-256 [2]

References

1. K. E. Murray, O. Petelin, S. Zhong, J. M. Wang, M. Eldafrawy, J.-P. Legault, E. Sha, A. G. Graham, J. Wu, M. J. P. Walker, H. Zeng, P. Patros, J. Luu, K. B. Kent, and V. Betz, "VTR 8: High-performance cad and customizable FPGA architecture modelling," *ACM Transactions on Reconfigurable Technology and Systems*, vol. 13, no. 2, 2020.
2. Z. U. Abideen, T. D. Perez, M. Martins, and S. Pagliarini, "A security-aware and lut-based cad flow for the physical synthesis of hasics," *IEEE Transactions on Computer-Aided Design of Integrated Circuits and Systems*, vol. 42, no. 10, pp. 3157–3170, 2023.
3. M. G. A. Martins, R. P. Ribas, and A. I. Reis, "Functional composition: A new paradigm for performing logic synthesis," in *Thirteenth International Symposium on Quality Electronic Design (ISQED)*, pp. 236–242, 2012.
4. M. G. A. Martins, L. Rosa, A. B. Rasmussen, R. P. Ribas, and A. I. Reis, "Boolean factoring with multi-objective goals," in *Computer Design (ICCD), 2010 IEEE International Conference on*, pp. 229–234, IEEE, 2010.
5. Z. U. Abideen, T. D. Perez, and S. Pagliarini, "From FPGAs to obfuscated eASICs: Design and security trade-offs," in *2021 Asian Hardware Oriented Security and Trust Symposium (AsianHOST)*, pp. 1–4, 2021.
6. M. Imran, Z. U. Abideen, and S. Pagliarini, "An open-source library of large integer polynomial multipliers," in *2021 24th International Symposium on Design and Diagnostics of Electronic Circuits Systems (DDECS)*, pp. 145–150, 2021.

7. FreeCores, "Infinite impulse response (IIR) filter," last accessed on Dec 25, 2021. Available at: https://github.com/freecores/all-pole_filters.
8. T. Zhu, "PID (proportional integral derivative) controller," last accessed on Dec 26, 2022. Available at: https://opencores.org/projects/pid_controller.
9. J. Carlos, "FPGA-based median filter," last accessed on Feb 19, 2023. Available at: https://opencores.org/projects/fpu100.
10. S. Joachim, "SHA-256," last accessed on Jan 20, 2023. Available at: https://github.com/secworks/sha256.
11. J. Al-Eryani, "Floating-point unit (FPU) controller," last accessed on Feb 15, 2023. Available at: https://opencores.org/projects/fpu100.
12. O. Kindgren and M. John, "OpenRISC 1200 implementation," last accessed on Feb 21, 2023. Available at: https://github.com/openrisc/or1200.

Physical Implementation of hASIC 10

10.1 Physical Synthesis Flow

This section discusses the validation of the hASIC design obfuscation technique in physical implementation. The aim is to analyze the impact of various obfuscation levels on physical synthesis results, including area, power, timing, and security trade-offs.

TOTe generates a structured Verilog representation of the hASIC when the design is obfuscated. The process of implementing an hASIC can be divided into two phases: logic synthesis and physical synthesis. Figure 10.1 provides a detailed diagram flow of this process. Logic synthesis converts the Verilog code into a gate-level netlist while meeting the performance requirements specified in the design constraints. Generating the gate-level netlist requires a standard cell IP library and design constraints. Therefore, the designer must have decided on the technology to fabricate the IC during this phase. The inputs required for the logic synthesis are the Verilog code of hASIC, the standard cell timing library, and design constraints.

Generally, foundries characterize each gate regarding process variation, voltage, and temperature in timing libraries. These characteristics are typically compiled in a standard Liberty file. In addition to the timing information, the designer must set the design constraints using the Synopsys Design Constraint (SDC) format [1]. The SDC file is where all clocks, input delay, output delay, and many other parameters can be described and constrained. Although the gate-level netlist is useful for estimating PPA values, physical synthesis provides more accurate results due to its consideration of routing, placement, and clock propagation.

As shown in Fig. 10.1, the output of physical synthesis is a layout used by the foundry as a blueprint for the fabrication of an IC, which is usually handled in GDSII format. At this level, physical information about standard cells and the metal stack is required. The EDA tool is able to manage metal layers, including the number of metals, the allowable width of each metal, and the type of vias used to connect the metals. The Library Exchange Format (LEF) file is the preferred format for describing the physical characteristics of each gate

Z. U. Abideen and S. Pagliarini, *Reconfigurable Obfuscation Techniques for the IC Supply Chain*, Synthesis Lectures on Digital Circuits & Systems,
https://doi.org/10.1007/978-3-031-77509-3_10

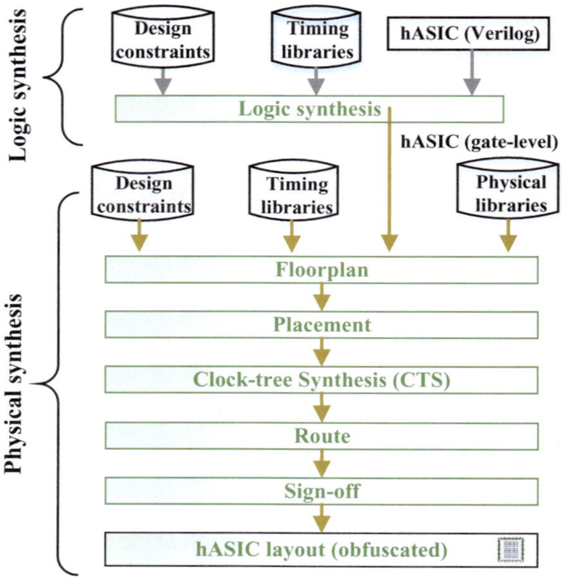

Fig. 10.1 The steps involved in the physical synthesis of hASIC [2, 3]

and the metal stack [4]. Inputs for physical synthesis include the gate-level netlist, timing libraries, design constraints, and LEF files for the cells and technology.

The physical synthesis process consists of several steps. Floorplanning involves sizing the design according to a target cell density, defining the pinout, and implementing power distribution. At this stage, the density setting is relatively precise since all the required gates except for the buffers in the netlist are present. The gates will still be modified while the optimization phase is ongoing. After floorplanning comes placement, which not only places gates coherently with their interconnections, but also considers timing and congestion.

The next step is the clock implementation that happens during CTS [5]. The main goal of CTS is to deliver the clock signal to all sequential elements in the design while balancing or skewing the clock accordingly. With all gates placed and the clock tree routed, the next step is to route all interconnected gates. Routing involves drawing wires between all drivers and sinks, i.e., the output and input pins of standard cells. Depending on the amount of routing resources, design rules, and congestion, routing can be very challenging, taking several hours or even days to complete. After routing, sign-off completes the design process for hASIC to carry out physical verification. Finally, the layout of hASIC is exported in GDSII format.

Physical synthesis is a complex and time-consuming process. We will consider two designs and present the final layouts for various obfuscation levels. These designs are well-known cryptocores, namely AES-128 [6] and SHA-256 [7]. These designs are medium-sized and represent practical examples. During physical implementation, Cadence Innovus is used with a commercial 65nm PDK for physical synthesis [8].

10.2 Physical Implementation of AES-128

Three distinct obfuscation levels, 60, 70, and 80%, were selected to analyze the design versus security trade-offs. LUTs defined in Sect. 9.3.1 are employed for the design with aforementioned obfuscation levels. Table 10.1 presents the results after the physical synthesis of AES-128. The analysis started with the initial FPGA implementation, which achieved a frequency of only 103 MHz, with the target device being a Kintex-7. Table 10.1 provides the results for obfuscation levels of 80%, 70%, and 60%. The results indicate that the level of obfuscation does not affect the utilization density of the design, which is determined by the ratio of placement sites that are occupied vs. total area. This indicates that the macros do not compromise global routing resources, i.e., hASIC is not limited by wire congestion as the obfuscation levels increase. The designs achieved a high utilization density of around 80% for all designs, even with the inclusion of many LUTs.

The implementation for the 60% obfuscation level shows the lowest area and a performance of 260 MHz for TOTe. The results also demonstrate that decreasing the obfuscation level improves the frequency, and vice versa. Timing results were obtained after physical synthesis and are for the worst process corner (SS), $VDD = 0.9 * VDD_{nominal}$, and a temperature of 125 °C. It is worth noting that the area of TOTe-generated designs increases as the obfuscation level increases, and the number of LUTs also increases with the obfuscation level. This behavior aligns with the original goal of TOTe, which was to establish a trade-off between performance (ASIC cells) and security (FPGA-like LUTs). As previously mentioned, LUTs are used for security purposes but exhibit an area penalty, as confirmed by physical synthesis. Furthermore, leakage and dynamic power values are proportional to security since reconfigurable logic is less efficient in terms of frequency than static logic.

Last, the last five columns of Table 10.1 display the resource utilization for hASIC, including the number of buffers, combinational cells, inverters, sequential cells, and the total wirelength. The total wirelength of a design is the combined length of all wires of all signals. Minimizing wirelength is one of the objectices of a physical synthesis tools since it helps to reduce the chip's size and cost. Additionally, minimizing the length of wires also reduces power consumption and delay, which are directly proportional to the wire length.

Table 10.1 Implementation results of AES-128 under different obfuscation levels

Design	Obf. (%)	Dens. (%)	Area (mm^2)	Freq. (MHz)	Leakage Power (mW)	Dynamic Power (mW)	# LUT	# Buffer	# Comb.	# Inv.	# Seq.	Total Wire-length (mm)
FPGA	100	–	–	103	6.2	587	10688	–	–	–	6000	–
TOTe	80	78	14.062	240	97.51	2246.49	9332	31376	7527	3634	15332	17432.17
TOTe	70	80	12.118	249	86.52	1989.45	8165	27972	14165	4440	14165	15758.92
TOTe	60	81	10.386	260	75.43	1744.57	6999	23398	28395	5491	12999	13846.95
ASIC	–	73	0.410	833	4.32	124.08	–	1810	99394	9769	6000	2664.06

Obf. is obfuscation, Dens. is density, comb. is combinational, inv. is inverters, and seq. is sequential

By examining the last column of Table 10.1, it is evident that the total wire length increases as the obfuscation level increases. The ASIC results show higher performance and lower PPA values. The targeted frequency is the maximum frequency. Therefore, the results for TOTe lie between FPGA and ASIC, as initially hypothesized.

In Fig. 10.2, different views of layouts of AES-128 under various obfuscation levels are presented. The metal stack considered here has seven metals assigned to signal routing (M1-M7). Figure 10.2a–c depict the layouts for 60, 70, and 80% obfuscation levels. The dimensions of layouts are included on the bottom and left sides of each panel. All six variants of LUTs are highlighted with different colors. The static part of hASIC is highlighted in red, and as expected, the design remains predominantly a sea of LUTs. The design comprises LUT_4 and LUT_6, but LUT_6 constitutes the majority. Therefore, the layouts seem to be dominated by orange boxes.

Figure 10.2d–f show the layouts for a 70% obfuscation level. Figure 10.2d illustrates the layout after routing. The design features mostly vertical orange lines corresponding to M6. Figure 10.2e provides a closer look at the placement pattern in an hASIC design, which consists of a combination of LUT macros and standard cells. The macros are positioned in alignment with the standard cell rows, leading to a uniform power rail and power stripe configuration throughout the design. The space between the macros is filled with standard cells. Figure 10.2f highlights the same design but with some routing layers filtered out (only M2, M3, and M4 are shown). As seen in Fig. 9.4, the implemented LUTs utilize the mentioned metal layers, resulting in a visually regular hASIC structure in Fig. 10.2f.

10.3 Physical Implementation of SHA-256

In the following subsection, the physical synthesis of SHA-256 has been performed with the same obfuscation level as mentioned earlier. Additionally, the physical synthesis of optimized designs is also presented for comparison. The Verilog code of the SHA256 core was obtained from an open-source repository [7]. Similar to the results for AES-128 on FPGA, the device achieved a frequency of 77 MHz for SHA-256. Table 10.2 shows the utilization of LUTs and FFs for SHA-256. Another point for analysis is being considered here, starting with the 100% obfuscation level as a baseline. This level is fully reconfigurable and comparable to an FPGA design. When compared to FPGA, hASIC exhibits higher performance. However, it also shows increased leakage, dynamic power, and the number of FFs. This FF count includes the ones required for configuring the bitstream as well as the ones needed for the design itself.

Let us consider the analysis with a 5% increment in obfuscation level. Table 10.2 shows the implementation results for obfuscation levels of 90, 85, 80%. The implementation of SHA-256 showed a similar trend to that observed in the initial analysis made by TOTe. Similar to AES-128, the utilization density of the design remained around 80% for all designs despite a large number of macros. Similar to AES-128, the trend is evident from

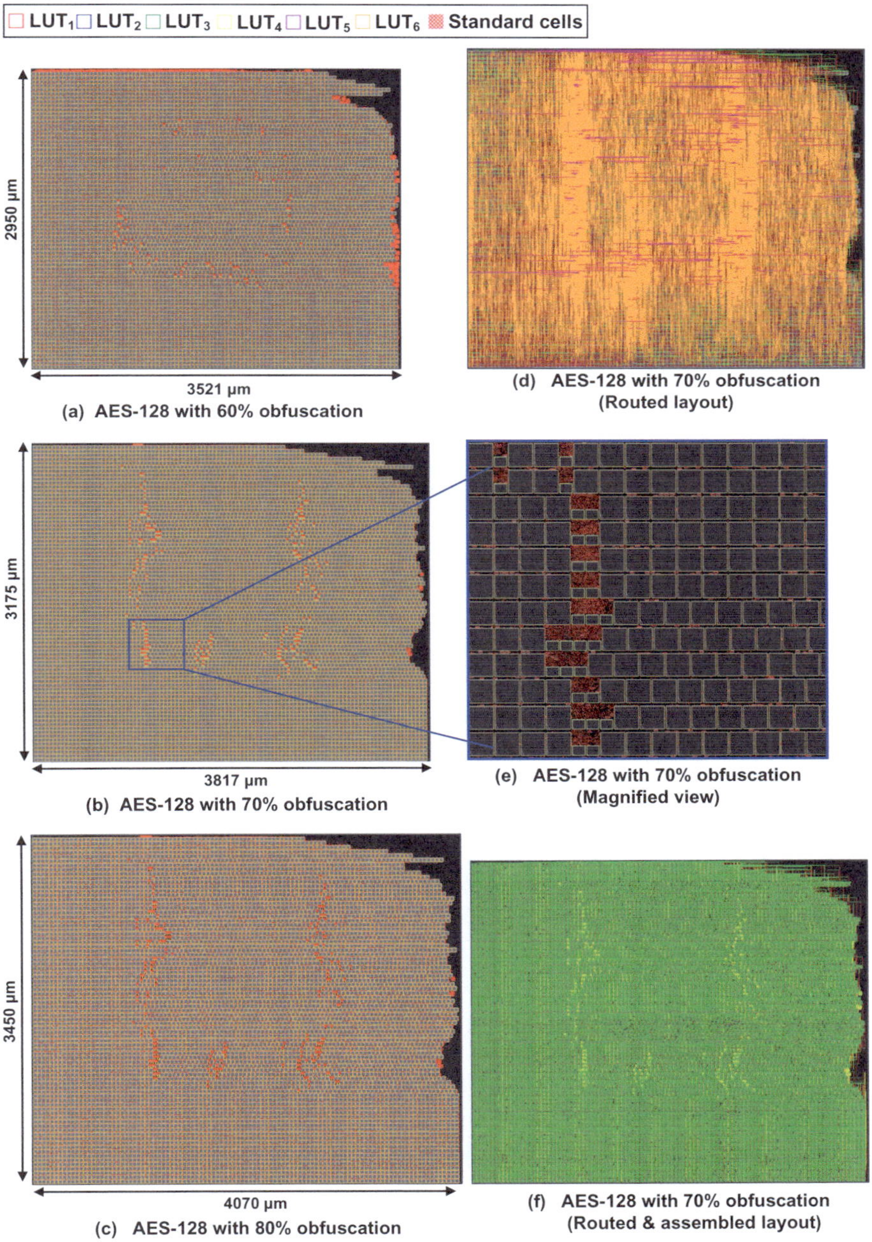

Fig. 10.2 Implementation results for AES-128 with different obfuscation levels

Table 10.2 Implementation results for baseline and optimized variants of SHA-256 under different obfuscation levels

Design	Opt.	Obf. (%)	Dens. (%)	Area (mm^2)	Freq. (MHz)	Leak. Power (mW)	Dyn. Power (mW)	# LUT	# Buffer	# Comb.	# Inv.	# Seq.	Total Wire-length (mm)
TOTe	No	100	81	1.751	223	14.85	505.05	2238	5846	93470	6175	105128	9247.65
TOTe	No	90	77	1.638	234	12.23	438.47	2015	4626	84107	5017	94876	7505.59
TOTe	No	85	80	1.507	241	12.10	430.98	1904	4846	80304	5585	90420	7207.02
TOTe	No	80	80	1.409	248	11.05	386.89	1792	4406	75083	4564	83790	6724.43
TOTe	Yes	100	61	1.155	307	8.00	301.49	10182	3583	29352	15261	53868	3391.74
TOTe	Yes	90	65	0.979	312	7.55	273.54	9127	1797	27115	13538	49016	3242.97
TOTe	Yes	85	67	0.940	322	7.03	256.36	8676	1882	26011	13136	46796	2982.62
TOTe	Yes	80	64	0.883	357	6.44	278.37	8124	1726	24614	12340	43830	2889.25
TOTe (Swap)	Yes	80	64	0.883	368	5.93	283.35	8124	1726	24614	12340	43830	2889.76
ASIC	–	–	92	0.034	248	0.18	9.37	–	167	3244	190	1806	158.00
ASIC	–	–	91	0.040	550	0.299	23.86	–	675	7981	1456	1806	181.44

Obf. is obfuscation, Dens. is density, comb. is combinational, inv. is inverters, and seq. is sequential

Table 10.2: Increasing the security of the design incurs PPA penalties and vice versa. The baseline hASIC design runs at 223 MHz, as shown in the Freq. column of Table 10.2. The leakage and dynamic power figures are proportional to security, as reconfigurable logic is less efficient than the static part in hASIC. Similarly, the number of LUTs also decreases as the obfuscation level decreases and vice versa. The last five columns of Table 10.2 show the resource requirements for hASIC (number of buffers, combinational cells, inverters, sequential cells, and the total wirelength). All these observations were also analyzed for AES-128.

Figure 10.3a–c illustrate the layouts for 80, 85 and 90% obfuscation levels. The dimensions of the layouts are indicated on the bottom and left sides of each panel. As expected, the design remains primarily a sea of LUTs. From visual inspection, the difference between AES-128 and SHA-256 is clear. AES-128 used LUT_4 and LUT_6 only, but the SHA-256 uses all different LUT variants. This is because the commercial synthesis tool prioritizes using large LUTs, such as LUT_6, to maximize their utilization. In contrast, the FPGA synthesis tool is targeted with LUT_4, which differs from the standard approach taken by most commercial tools during synthesis. During placement, this technique aids in creating a more streamlined and uniform structure for the hASIC, resulting in a more compact design. All six variants of LUTs are highlighted with different colors and the static part of hASIC is highlighted in red. Figure 10.3d demonstrates the final post route layout of hASIC under 85% obfuscation level. Figure 10.3e shows the magnified view of the placement in an hASIC design under 85% obfuscation level. Figure 10.3f illustrates the assembled view of design under 85% obfuscation level with certain routing layers filtered out. Only M2, M3, and M4 are shown.

The same levels of obfuscation were taken into consideration when working on the optimized designs that leverage the decomposition and pin swap approaches previously

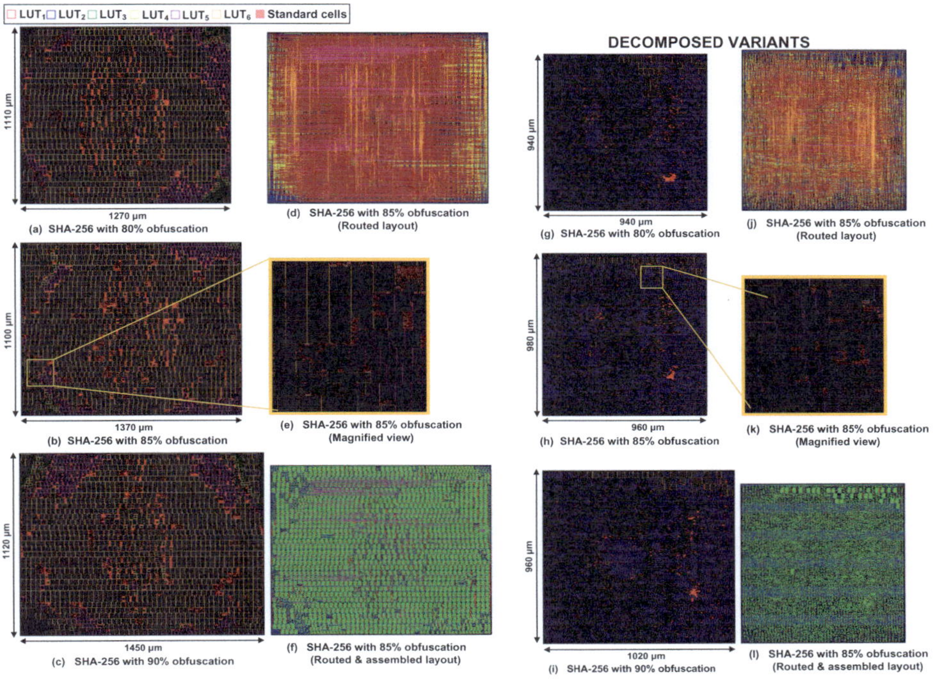

Fig. 10.3 Implementation results for SHA-256 with different obfuscation levels [3]

discussed. The analysis in Table 10.2 shows a similar trend for the optimized designs. With optimization, the baseline frequency increased significantly from 223 to 307 MHz. However, placing and routing become more challenging due to a large number of small LUTs, mainly LUT_2s. As a result, the maximum utilization density is approximately 65% across optimized designs. The optimized design resulted in an area reduction of 36% compared to baseline designs. Regarding frequency improvement, designs with obfuscation levels of 100, 90, and 85% saw a 35% improvement on average. However, the design at 80% obfuscation showed a frequency improvement of 43%.

To enhance the performance, the next step is to apply the pin-swapping technique. In this technique, the same logic function can be generated using different input orders and masking bits (truth table). This technique can swap their pins and effectively reduce the overall delay by identifying LUTs that appear on the critical paths. To demonstrate the effectiveness of pin swapping, a hypothetical situation is considered, where the target frequency of the design is increased, resulting in several paths violating setup timing. The number of violating paths determined the number of LUTs considered for pin swapping, establishing a trade-off between runtime and quality of results. More aggressive frequency targets meant more LUTs were considered for pin swapping. All LUTs from the violating paths were selected as candidates and saved in a list. Starting with the worst violating path, the pins of the LUTs

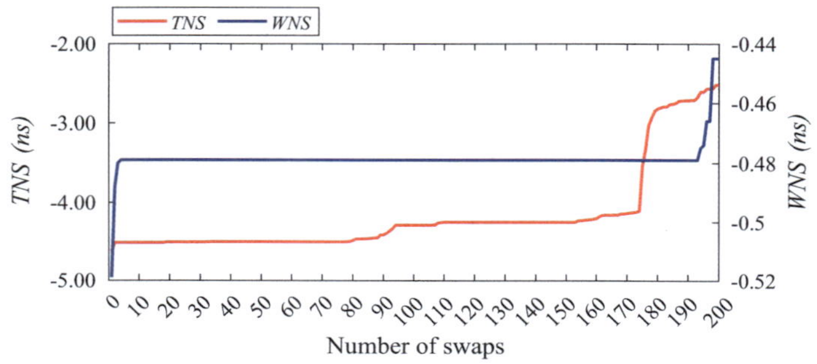

Fig. 10.4 Change in the TNS/WNS concerning the swap of LUT pins [3]

were iteratively swapped until the critical path was improved, as measured by WNS. The number of swaps versus TNS and WNS is shown in Fig. 10.4. The initial swaps improved WNS without any effect on TNS. However, continuing to swap improved the TNS without any change in WNS. Improving TNS indicates a potential for a better WNS, so swapping continued until the next jump in WNS. After 200 swaps, WNS improved by approximately 70ps, and TNS improved by 2ns, increasing the frequency of design by 11 MHz. With an obfuscation level of 80%, the same design exhibited a 48% performance improvement compared to the baseline design. It is clear that decomposition is highly beneficial, offering significant PPA gains compared to non-optimized versions. Pin swapping, while effective, produces more modest benefits.

The runtime of the physical synthesis flow is not significantly impacted by decomposition, making it worthwhile for design optimization. For instance, the runtime to apply the decompositions in the SHA-256 circuit with 100% obfuscation level containing 2238 LUTs was 11 minutes on an Intel Core i7-6700K. When obfuscation levels were applied at 100, 90, and 85%, the designs resulted in an average of 40% improvement in dynamic power. But, when the 80% obfuscation level is considered, only a 27% improvement is shown.

Figure 10.3g–l presents the layouts of optimized designs while maintaining the same design and conditions. The scaled layouts highlight the area reduction achieved by decomposition. Figure 10.3l still demonstrates regularity after decomposition, which is welcomed and expected. Regarding the ASIC implementation, the best possible frequency is 550 MHz. Area, leakage, dynamic power and other values are listed in Table 10.2. The number of FFs in the FPGA implementation differs from the ASIC implementation due to register duplication, a technique that synthesis tools often employ. From Table 10.2, it is evident that Vivado uses FF cloning to address high fanout buffering for SHA-256. Thus, there is an increase in the number of registers with respect to ASIC.

It is worth noting that hASIC has a highly regular structure upon visual inspection. This characteristic can be adjusted to enhance its effectiveness against reverse engineering adversaries. One way to achieve this is by mapping LUTs of various sizes to LUT_6, creating an even more uniform layout. Another option is to arrange LUTs in a perfect grid pattern. While both design choices are relatively straightforward to implement during physical synthesis, they also come with additional overhead costs that may not be beneficial.

In recent obfuscation research, there has been a growing trend in using eFPGA technology [9, 10]. This approach offers several advantages but is typically employed selectively to protect only specific design parts, thereby minimizing the performance penalty. However, the challenge lies in determining which circuit modules require protection and which do not. The methodology of hASIC is designed to bypass this issue by only revealing (parts of) critical paths as they are selected for static logic. This approach offers a distinct advantage over other methods. The authors of [11] present a top-down methodology for implementing ASICs with eFPGAs. Their designs share many similarities with hASIC solution while incorporating more regularity through logic tiles, similar to those found in commercial FPGAs. hASIC, which does not utilize tiles, prioritizes performance, as shown in the layouts of Fig. 10.3 and the corresponding results in Table 10.2. Our understanding is that IP protection, i.e., obfuscation, cannot come at the cost of performance. It is for this reason that hASIC focuses on attaining an ASIC-like level of performance and does so successfully.

References

1. Intel Inc., "Synopsys design constraints file (.sdc) definition." https://www.intel.com/content/www/us/en/programmable/quartushelp/17.0/reference/glossary/def_sdc.htm, 2024. Accessed: August 16, 2024.
2. Z. U. Abideen, S. Gokulanathan, M. J. Aljafar, and S. Pagliarini,"2021 Asian Hardware Oriented Security and Trust Symposium (AsianHOST)," *From FPGAs to Obfuscated eASICs: Design and Security Trade-offs*, p. (1–4) 2021.
3. Z. U. Abideen, T. D. Perez, M. Martins, and S. Pagliarini, "A security-aware and lut-based cad flow for the physical synthesis of hasics," *IEEE Transactions on Computer-Aided Design of Integrated Circuits and Systems*, vol. 42, no. 10, pp. 3157–3170, 2023.
4. Cadence Design Systems, Inc., "Lef/def 5.8 language reference." https://coriolis.lip6.fr/doc/lefdef/lefdefref/LEFSyntax.html, 2024. Accessed: August 16, 2024.
5. K. Golshan, *Clock Tree Synthesis*, pp. 123–137. Cham: Springer International Publishing, 2020.
6. H. Hsing, "AES-128," last accessed on Jan 22, 2023. Available at: https://opencores.org/projects/tiny_aes.
7. S. Joachim, "SHA-256," last accessed on Jan 20, 2023. Available at: https://github.com/secworks/sha256.
8. TSMC, Ltd, "65nm technology." https://www.tsmc.com/english/dedicatedFoundry/technology/logic/l_65nm, 2024. Accessed: August 16, 2024.
9. B. Hu, T. Jingxiang, S. Mustafa, R. R. Gaurav, S. William, M. Yiorgos, C. S. Benjamin, and S. Carl, "Functional obfuscation of hardware accelerators through selective partial design extraction onto an embedded FPGA," in *Proceedings of the 2019 Great Lakes Symposium on VLSI*, p. 171–176, 2019.
10. J. Chen, M. Zaman, Y. Makris, R. D. S. Blanton, S. Mitra, and B. C. Schafer, "DECOY: DEflection-Driven HLS-Based Computation Partitioning for Obfuscating Intellectual PropertY," in *Proceedings of the 57th ACM/EDAC/IEEE Design Automation Conference*, DAC '20, IEEE Press, 2020.

11. P. Mohan, O. Atli, O. Kibar, M. Zackriya, L. Pileggi, and K. Mai, "Top-down physical design of soft embedded FPGA fabrics," in *The 2021 ACM/SIGDA International Symposium on Field-Programmable Gate Arrays*, p. 1–10, 2021.

Security Analysis for hASIC 11

11.1 Revised Threat Model

When considering the threat model for hASIC, the main concern is the untrusted foundry, regardless of whether the adversary is an institutional entity or a rogue employee. The security of the design depends on both the static part, which is fully exposed, and the reconfigurable part, which is protected by a bitstream serving as key. In an eFPGA-based solution, it is likely that the entire sensitive part of the design is obfuscated inside the eFPGA macro. This is no longer true for hASIC, which then requires us to revisit the threat model and the adversarial capabilities.

The static part is, unfortunately, a source of vulnerability as the adversary can extract relevant information from it. An adversary can exploit the regular structure of a design (e.g., AES-128) to extract valuable information, thereby making it possible for the adversary to perform and educated guess of the bitstream. Furthermore, the adversary can easily guess the bitstream of the reconfigurable part if the static part is too large. On the other hand, if the reconfigurable part is too large, it provides high security, but leads to overheads in terms of PPA. Therefore, it is important to determine the *right level* of obfuscation to ensure that the design is secure against well-known attacks. Based on these factors, the following assumptions are considered for the hASIC-specific threat model:

- The adversary aims to reverse engineer the design to copy its IP, produce excessive amounts of the IC, or potentially implant complex hardware trojans. To accomplish this, the adversary is required to discover the bitstream.
- The adversary may have the objective of determining the circuit's functionality, even if obfuscation techniques have been implemented. It is worth noting that in this scenario, the adversary does not necessarily need to discover the bitstream.

Z. U. Abideen and S. Pagliarini, *Reconfigurable Obfuscation Techniques for the IC Supply Chain*, Synthesis Lectures on Digital Circuits & Systems,
https://doi.org/10.1007/978-3-031-77509-3_11

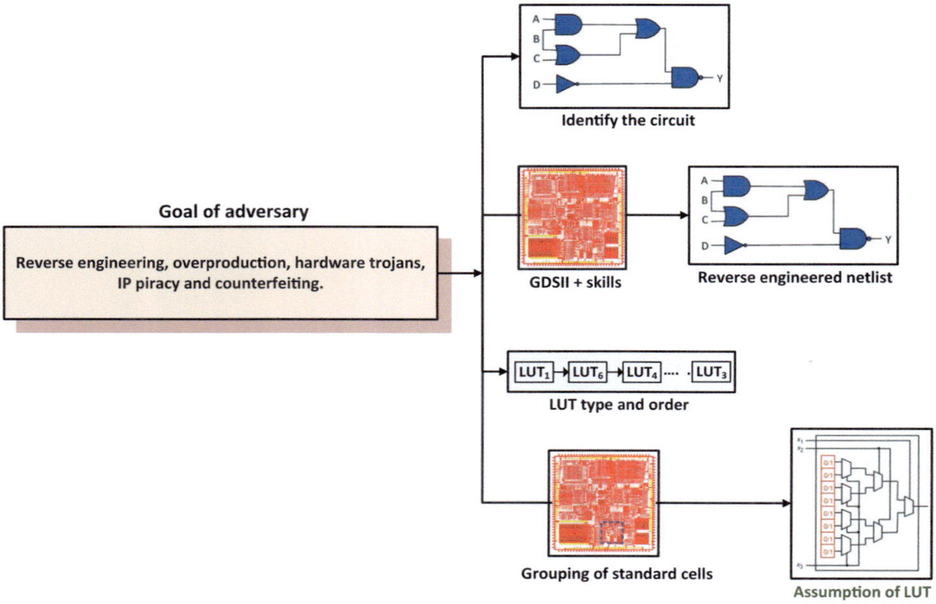

Fig. 11.1 The summary of the threat model for hASIC

- Due to their proficiency in IC design, the adversary possesses the necessary expertise and resources/tools to comprehend the layout. They have access to the GDSII file of the hASIC design submitted for fabrication.
- The adversary can identify the standard cells. Consequently, the gate-level netlist of the obfuscated circuit can be retrieved without much difficulty [1].
- Through analyzing the reconfiguration pins, the adversary can easily identify all LUTs and their programming order with no difficulty [2, 3].
- When obfuscating a design, the process starts with an FPGA and then moves to hASIC. hASIC is like an FPGA at the beginning, where the design is implemented. Assuming a perfect reconstruction of LUTs, the adversary can group the standard cells found within the static logic and convert them back to an LUT representation.

A summary of the threat model is depicted in Fig. 11.1.

An assessment was conducted to evaluate the security resistance of hASIC against conventional oracle-guided and oracle-less attacks, which are commonly used against LL. All the experiments were performed on a server with 32 processors (Intel(R) Xeon(R) Platinum 8356H CPU @ 3.90GHz) and 1.48TB of RAM. In order to assess the security of hASIC oracle-guided attacks (such as the SAT attack) and oracle-less attacks, security evaluation techniques like Synthesis-based COnstant Propagation Attack for Security Evaluation (SCOPE), as well as custom structural and compositional analysis attacks are utilized.

11.2 Oracle-Guided Attacks

The goal of an oracle-guided attack is to retrieve a key or a key guess. In hASIC, LUTs act as key gates in contrast to the traditional LL [4]. A single LUT_6 provides a 64-bit long key for obfuscation in hASIC. The SBM circuit, previously presented in Sect. 9.4, has a total of 25 LUTs, including 11 LUT_6, when obfuscated at a rate of 86%. The LUT_6s alone contribute to a key search space of $2^{11\times64}$, which is extremely discouraging for an adversary attempting SAT attacks on hASIC. However, enumerating the key search space may seem simple, but it is a naive approach to evaluate security. Actual attacks, particularly well-known SAT attacks, are necessary. Three different SAT attacks are employed to evaluate the security hardness of hASIC. These attacks are conventional SAT [5], AppSAT [6], and ATPG-based SAT [7]. These attacks operate on bench files and accept only combinational circuits as input. hASIC uses FFs to store a serial input bitstream. A script is written to convert them to combinational logic with parallel key bits. This is a relatively simple procedure that can be assumed easy for an attacker to execute.

Next, we have selected large combinational circuits, *c*6288 and *c*7552, to evaluate hASIC's security against SAT attacks. The results for the selected designs are presented for two different variants of hASIC, baseline and optimized. Figure 11.2a and b show the execution time for different SAT attacks at varying obfuscation rates for c7552 and c6288, respectively. As expected, the execution time increases with the increase in obfuscation level. The region to the left of the green line displays successful SAT attacks, while the region on the right corresponds to unsuccessful attacks where the solver took more than 48 hours to return an answer. Different designs can experience timeouts at varying obfuscation rates

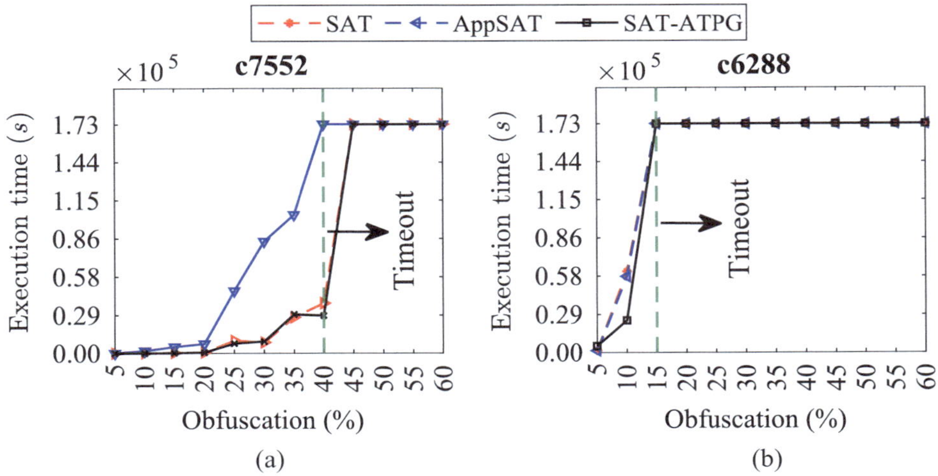

Fig. 11.2 The execution time of SAT attacks

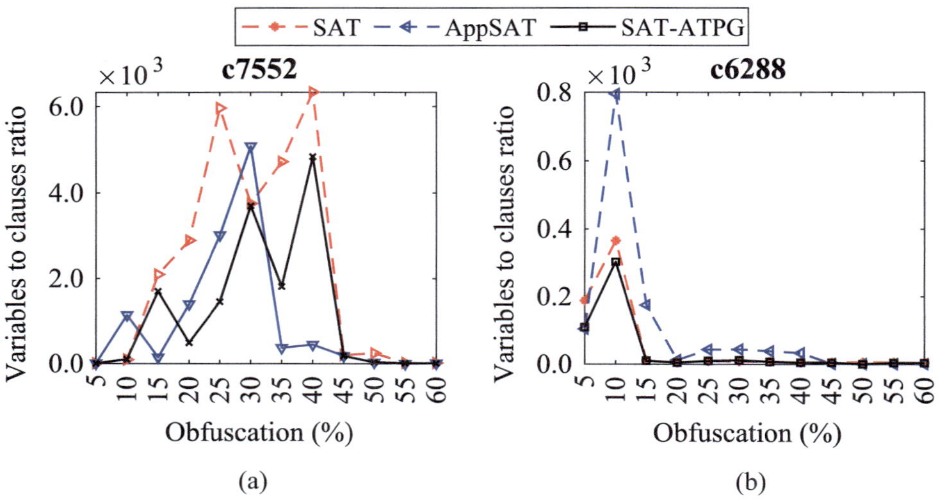

Fig. 11.3 The variables to clauses ratio of SAT attacks for two different designs

when using different SAT solvers. For $c7552$, we encountered timeouts at a 40% obfuscation rate, while $c6288$ already experienced them at a 15% obfuscation rate. It is important to note that these designs are extremely small by modern standards, and yet appear to resist to SAT attacks convincingly.

In this analysis, the SAT problem is considered as a *circuitSAT* problem in order to understand better the behavior of the SAT solver for the selected designs. The analysis is customized specifically for the selected designs. Any other benchmarks may require a different analysis. Figure 11.3 illustrates the behavior of the SAT solver when dealing with obfuscated circuits. The SAT solver is responsible for determining the satisfiability of a boolean formula. One way to measure the attack convergence probability is by calculating the ratio of variables-to-clauses of the SAT solver. Figures 11.3a and b show the progression of the variables to clauses ratio for $c7552$ and $c6288$, respectively, as the obfuscation level increases. The lower the value of the ratio, the more complex the *circuitSAT* problem becomes. The complexity trend varies with the obfuscation level, but the problem becomes hard after 45% obfuscation level for $c7552$. On the other hand, the problem becomes difficult after 20% obfuscation level for $c6288$. These findings are in alignment with the timeout profiles previously described.

To enhance power, area, and performance, TOTe was given the capability to decompose LUTs into smaller ones, resulting in improved QoR. However, this process also reduces the size of the bitstream, *potentially* making it vulnerable to attacks. To ensure the security of the optimized version of the $c7552$ design, it is necessary to verify that the reduced bitstream size does not expose it to existing attacks. Figure 11.4a shows the execution time for the optimized variant of $c7552$, while Fig. 11.4b displays the variable to clauses ratio.

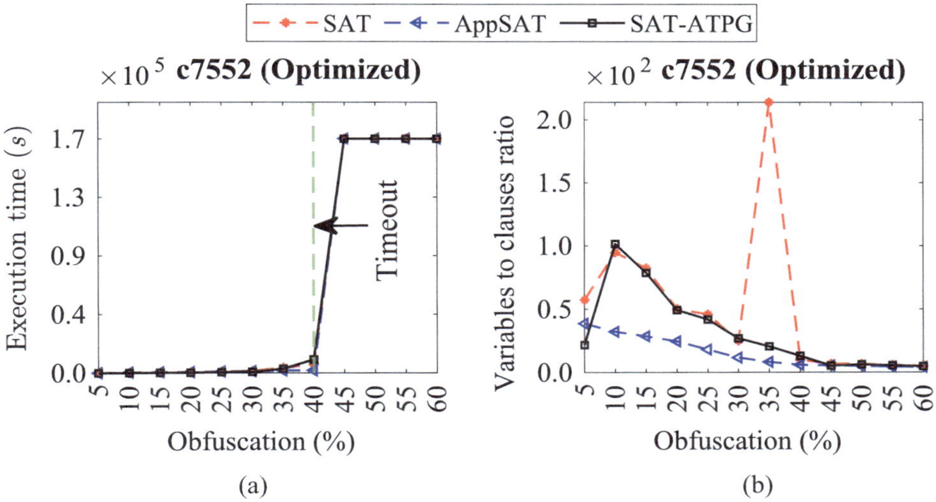

Fig. 11.4 The optimized results for $c7552$ regarding the execution time and the ratio of variables to clauses in SAT attacks

The security analysis indicates that successful attacks take less time to complete, but none succeed beyond 40% obfuscation. Additional information on the $c7552$ design, including bitstream size for different designs and two obfuscation rates (55% and 60%), can be found in Table 11.1. Interestingly, the decomposed designs exhibit a better variables to clauses ratio, suggesting that the decomposition keeps keys less correlated with each other, making each individual key bit relatively more effective. However, this does not imply that the baseline designs were less secure. The analysis showed that the circuitSAT problem is hard for selected designs, considering different obfuscation levels and lower ratios. For more information on the SAT attack, please refer to [8], and for a discussion on key to key interference, see [7].

11.3 Oracle-Less Attacks

This section discusses oracle-less attacks including the SCOPE attack and custom attacks developed specifically for hASIC security analysis. To evaluate the security strength of hASIC, two different attacks have been proposed, one based on the design's structure and the other based on the composition of various circuits. It is believed that knowledge can be gained and information can be extracted by exploiting the static portion of the design, which includes the frequency of specific LUT masking patterns. Such capability would enable the attacker to reduce the search space for the key that unlocks the design. The frequency of masking patterns can be effectively used as a comparative template for assessing different

Table 11.1 Analysis of variables to clauses ratio for $obf_c = 55\%$ and 60%

Attack	Obf. (%)	*Bitstream length* (bits)	*Variables*	Clauses	Iterations	Ratio
SAT [5]	55	7494	32318070	1597559	820	17.2
	60	8582	33315150	1898618	540	17.5
	55∗	4014	3434578	598599	139	5.7
	60∗	4406	2829008	564533	106	5.0
AppSAT [6]	55	7994	15843074	1127281	8	14.0
	60	8582	15412584	1181330	7	13.0
	55∗	4014	1320652	302801	2	4.36
	60∗	4406	1419176	346847	2	4.09
ATPG-SAT [7]	55	7494	27243806	1800999	630	15.1
	60	8582	31166810	2247522	636	13.8
	55∗	4014	3642116	664817	148	5.4
	60∗	4406	2274084	480548	85	4.7

∗ Results for the optimized designs

designs. This means that the composition of LUTs in a design can be vulnerable to structural attacks [9].

11.3.1 SCOPE Attack

The SCOPE attack aims to retrieve a key but its heuristic nature can provide a key guess that is not entirely accurate. Oracle-less attacks refer to attacks that do not rely on a functional IC (oracle). Instead, they target the netlist of obfuscated circuits directly and infer key guesses from it. This attack requires no prior knowledge of obfuscation techniques. SCOPE analyzes a single key bit through (re-)synthesis and extracts crucial design features, such as area, power, and delay, that may enable the derivation of the correct key bits. Figure 11.5a compares the execution time for both the baseline and optimized designs of *c*7552. As shown in Fig. 11.5a, the execution time increases with the level of obfuscation. This trend is observed for both the baseline and optimized designs, with different rates of increase. It is important to note that the COPE metric provides a rough estimate of the level of vulnerability (%) to SCOPE attacks. Figure 11.5b shows the COPE metric, which decreases with increasing levels of obfuscation. The SCOPE metric should be lower, highlighting that the design is less vulnerable to attacks, as illustrated for a 60% obfuscation level in Fig. 11.5b.

After running SCOPE, a guess is generated for the key with each bit assigned either a ‘1’, a ‘0’, or an ‘X’ to indicate that it is undetermined. After comparing the guess generated

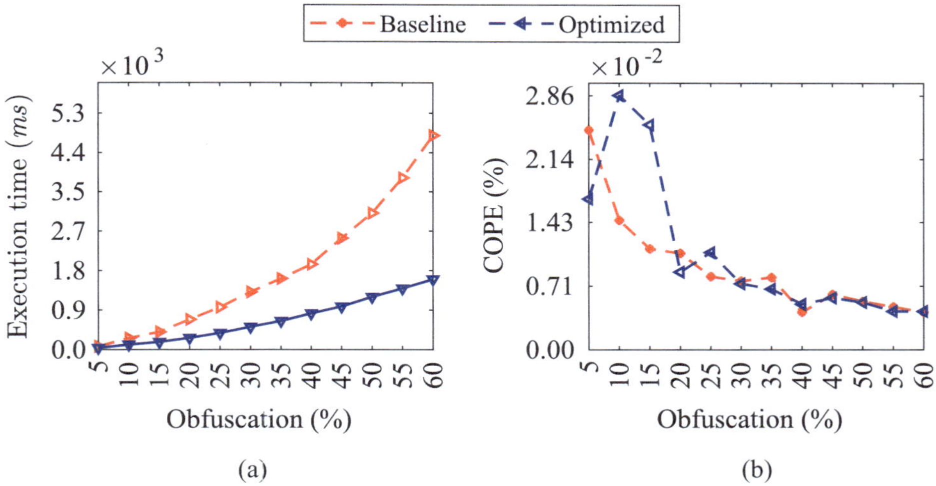

Fig. 11.5 The comparison between the baseline and optimized design for the oracle-less SCOPE attack

by SCOPE to the known key bits, it was discovered that 50% of the key bits were correctly guessed, which is a random guess regardless of the level of obfuscation. This percentage remains constant for both baseline and optimized designs. Therefore, for hASIC, it appears that SCOPE cannot perform better than a random guess.

To identify design intent, a composition analysis attack must have access to a high-quality database of known designs. Therefore, while using TOTe, it is recommended to employ very high obfuscation rates to prevent such attacks. Additionally, reconfigurable-based obfuscation schemes are generally less susceptible to attacks than LL counterparts, making it important to maintain high obfuscation rates.

11.3.2 Structural Analysis Attack

This attack aims to utilize statistical analysis methods to reduce the search space and facilitate the bitstream recovery process. The obfuscation engine used by TOTe consists of six variants of LUTs, with LUT_6 being the most prevalent due to the packing algorithm of commercial FPGA synthesis [10]. The decomposition method only applies to the reconfigurable part of the design, whereas the initial knowledge obtained by the adversary is from the static part, which remains unchanged regardless of the use of decomposition. The majority of the design is composed of LUT_6s, prompting the adversary to focus their analysis on this type of LUT. The number of possible keys for a LUT_6 is 2^{64}, but this is only feasible if the FPGA synthesis tool can realistically explore the entire key search space. However, it appears that this is not the case. All unique LUT_6 masking patterns were extracted from the netlists of

31 representative designs of varying size, complexity, and functionality through synthesis. These masking patterns are denoted as ump_i. The results of the analysis for various designs are presented in Table 11.2. From the third to the eighth column, the table shows the total number of corresponding LUTs and the unique making patterns for LUT_4, LUT_5, and LUT_6. The last three columns of this table illustrate the maximum frequency of the unique masking pattern in the listed design. After analyzing all the unique masking patterns, it was found

Table 11.2 Global search space analysis

Design	Obf. (%)	# LUT_6		# LUT_5		# LUT_4		Max. Frequency		
		Total	Unique	Total	Unique	Total	Unique	# LUT_6	# LUT_5	# LUT_4
c17 [11]	100	0	0	0	0	2	2	0	0	2
c432 [11]	100	20	19	49	36	92	38	3	7	19
c499 [11]	100	40	6	80	12	173	52	17	17	33
c880 [11]	100	34	27	87	57	143	52	5	9	25
c1355 [11]	100	18	3	62	10	88	6	11	23	49
c1908 [11]	100	33	28	89	67	148	59	4	6	22
c2670 [11]	100	51	25	119	49	226	66	17	18	17
c3540 [11]	100	133	86	323	181	570	206	14	27	80
c5315 [11]	100	143	90	333	181	611	211	10	15	39
c6288 [11]	100	419	45	847	94	1760	108	77	78	153
c7552 [11]	100	161	142	378	276	703	281	4	13	57
DES [12]	100	768	92	1593	151	3082	184	114	142	310
RSA [13]	100	586	115	1374	208	2477	177	172	172	472
GFX430 [14]	100	1212	519	3275	951	5149	676	85	193	671
MIPS [15]	100	3162	776	7124	662	13228	488	670	734	1372
JPEG DEC [16]	100	2413	347	5971	619	10585	505	401	510	1090
USB HOST [17]	100	502	123	1138	194	2067	175	264	215	528
CORDIC [18]	100	516	209	1141	330	2175	244	46	185	613
FM [19]	100	188	149	579	311	788	405	9	76	38
SIGMA DELTA [20]	100	32	3	66	7	218	12	29	30	61
openMSP430 [21]	100	760	371	2048	703	3624	509	26	142	439
SBM [22]	100	11	5	26	14	52	20	8	7	15
AES-128[23]	100	9280	45	0	0	1408	178	1153	0	209
SHAKE-256 [24]	100	4438	35	3083	64	5496	53	1395	1215	2584
PID [25]	100	364	175	989	336	1561	317	28	84	50
Median Filter [26]	100	410	27	1077	60	1815	60	97	129	407
SHA-256 [27]	100	1349	69	706	133	96	19	513	50	2584
GPU (OR1200-HP) [28]	100	22611	260	7563	458	758	374	11809	1236	618
RISC-V [29]	100	2240	628	5016	507	9831	404	508	300	1018
FPU [30]	100	823	233	2122	423	3764	373	48	70	372
IIR [31]	100	1	1	4	4	148	3	2	2	2
Total	–	52718	4653	47262	7098	72838	6257	–	–	–

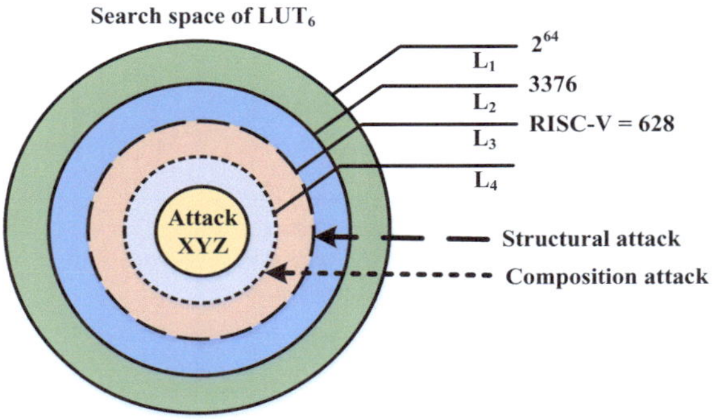

Fig. 11.6 The search space of LUT_6 as it shrinks with different attacks [32]

that the combined number of unique masking patterns for LUT_6 in the fourth column of Table 11.2 formed a set of $M = \bigcap_{i=1}^{31} |\{ump_i\}| = 3376$ elements, which appears to have settled. As shown in Fig. 11.6, this empirical result reduces the global search space from 2^{64} to $3376 = 2^{11.72}$.

Based on the available information, it appears that an attacker could exploit the frequency of LUTs within a netlist to launch a structural analysis attack. In order to do so, the attacker would need to determine the values of ump_i for a given circuit C_i, despite only having partial knowledge of the design. The question then becomes whether it is possible to estimate ump_i through statistical analysis of a part of C_i. To investigate, two processor designs were analyzed: MIPS and RISC-V. For each circuit, $\langle pattern, frequency \rangle$ tuples were used to track the repetition of masking patterns, with the masking pattern represented by 64-bit hexadecimal numbers and ordered by frequency. Figure 11.7a and b show the bar charts of RISC-V and MIPS, respectively. The MIPS netlist contains 776 unique LUT_6s, with only a few masking patterns that occur more than 50 times. Similarly, in RISC-V, there are 628 unique LUT_6s, with only three occurring more than 100 times.

Figure 11.8a and b investigate the masking pattern frequency for RISC-V and MIPS, respectively. Netlists generated by TOTe at different obfuscation levels were used for this purpose. The experiment assumes that the attacker has visibility of the static part with only a

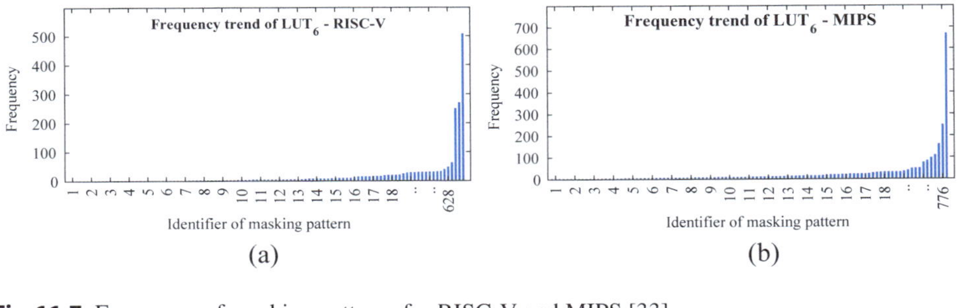

Fig. 11.7 Frequency of masking patterns for RISC-V and MIPS [33]

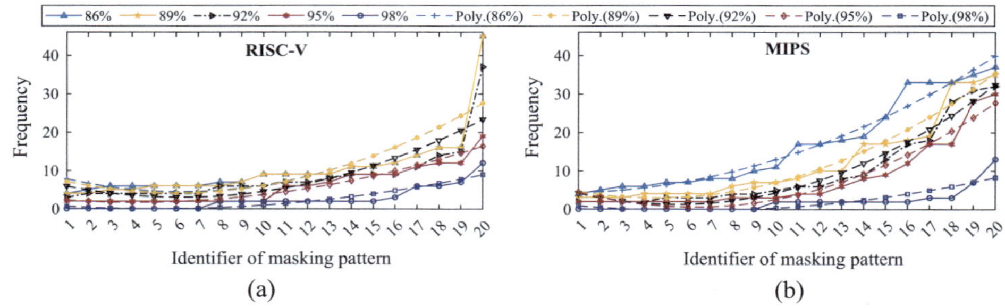

Fig. 11.8 The structural analysis of RISC-V and MIPS [33]

small percentage of LUTs, ranging from 2 to 14%, depending on the obfuscation level. The attacker then tries to predict the distribution of actual masking patterns in the entire design based on their observation of the exposed LUTs in the static portion of hASIC. Polynomial trendlines are used to aid the adversary's guessing attempt in Fig. 11.8. For MIPS and RISC-V, the attacker can estimate to some degree which masking patterns are unique. Extrapolation is not trivial to determine the actual number of unique masking patterns ump_i since many patterns appear only once or a few times, as shown in Fig. 11.7a and b. Some circuits, such as PID, IIR, GPU, and SHA-256 have a similar profile where only a few high-frequency LUTs appear. The attack exploits only the static part, but when the decomposition is applied, the adversary needs in-depth knowledge of the decomposition algorithm to estimate the frequency of the unique masking pattern for the reconfigurable part. In other words, here decomposition is likely to improve the security of the design.

11.3.3 Composition Analysis Attack

This attack aims to correlate an unknown circuit under attack with known circuits. The adversary's sole objective is deciphering a circuit (specifically, "What is this circuit?"). In this attack, the frequency of the LUTs is also exploited. However, it involves correlating multiple designs with one another based on their composition. It is worth noting that the attack can be deemed successful if the adversary can identify the circuit, rendering the need to break the key unnecessary.

Figure 11.9a and b show correlation analysis for two crypto cores: SHA-256 and AES-128, respectively. The goal of this analysis is to examine the leaked information from the static part against a database[1] of circuits that are known to the attacker. The obfuscation of SHA-256 and AES-128 is performed in the range of 70–100%, followed by the correlation of their static parts with the designs available in the database.

---

[1] It is assumed that the adversary can obtain circuits from open-source repositories and execute FPGA synthesis to create a relatively large database.

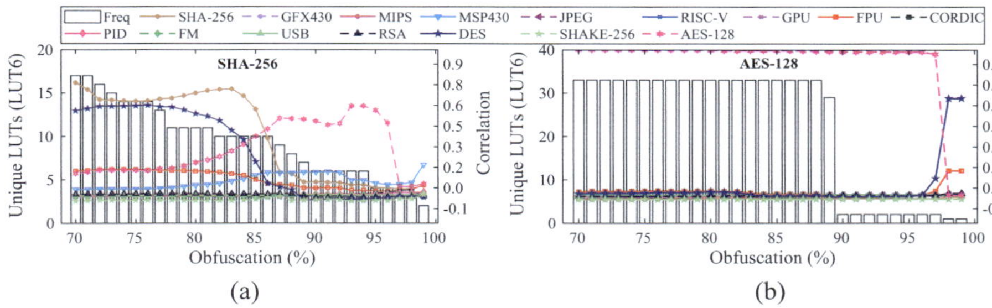

Fig. 11.9 The correlation of SHA-256 and AES-128 versus numerous other designs [33]

Based on the correlation results, interesting trends have been revealed. For SHA-256, in Fig. 11.9a, three regions of interest have been identified depending on the degree of obfuscation: 97–100% (no correlation), 86–96% (strong correlation to another circuit), and 70–85% (correlation to itself). Figure 11.9b shows a similar analysis for AES-128. The correlation between obfuscated AES-128 and itself is almost one for obfuscation levels below 97%, while the correlation for obfuscated AES-128 versus other designs is almost zero for obfuscation levels in the same range. In the scenario where the adversary can identify this potential vulnerability of hASIC, it could be equivalent to that of finding the intent of ASIC circuit. However, this is not the case for SHA-256, as opposed to AES-128, where different ranges of obfuscation levels can confuse the adversary. For SHA-256, this range is found to be between 86 and 96%. In this case, the search space will be shifted to L_4 for the corresponding design, as shown in Fig. 11.6.

To further reduce the key search space, an adversary interested in obtaining the bitstream could use the correlation analysis as follows. If the attacker knows that the obfuscated circuits are AES-128, SHA-256, or any other, his/her key guessing will rely on the circuit with the highest correlation. It is important to note that this attack depends on the adversary's ability to reconstruct the LUTs from the static part, and the availability of enough datapoints in the database of known circuits. For example, in the previous subsection, the search space would shrink from 3376 to 776 for MIPS and from 3376 to 628 for RISC-V. As mentioned in Sect. 11.3.2, this attack only exploits the static part of the design, and a decomposed design does not make it easier to correlate against a database. However, to obtain the actual key, an adversary would need to use other attacks which are not specific to hASIC. These attacks could be either an oracle-guided attack or any future 'XYZ' attack, as illustrated in Fig. 11.6.

It is worth noting that hASIC has a highly regular structure upon visual inspection. This characteristic can be adjusted to enhance its effectiveness against reverse engineering. One way to achieve this is by mapping LUTs of various sizes to LUT_6, creating a more uniform layout. Another option is to arrange LUTs in a perfect grid pattern. While both design choices are relatively straightforward to implement during physical synthesis, they also come with additional overhead costs that may not be beneficial. The choice of LUT regularity has little to no effect to the custom attacks described in this chapter. Furthermore, it would not prevent any SAT attack from being mounted. It may, however, prevent reverse engineering attacks that leverage structural traces.

References

1. R. S. Rajarathnam, Y. Lin, Y. Jin, and D. Z. Pan, "ReGDS: A reverse engineering framework from GDSII to gate-level netlist," in *2020 IEEE International Symposium on Hardware Oriented Security and Trust (HOST)*, pp. 154–163, 2020.
2. P. Subramanyan, S. Ray, and S. Malik, "Evaluating the security of logic encryption algorithms," in *2015 IEEE International Symposium on Hardware Oriented Security and Trust (HOST)*, pp. 137–143, 2015.
3. M. Yasin, A. Sengupta, M. T. Nabeel, M. Ashraf, J. J. Rajendran, and O. Sinanoglu, "Provably-secure logic locking: From theory to practice," in *Proceedings of the 2017 ACM SIGSAC Conference on Computer and Communications Security*, p. 1601–1618, 2017.
4. J. A. Roy, F. Koushanfar, and I. L. Markov, "EPIC: Ending piracy of integrated circuits," in *2008 Design, Automation and Test in Europe*, pp. 1069–1074, 2008.
5. P. Subramanyan, S. Ray, and S. Malik, "Evaluating the security of logic encryption algorithms," in *2015 IEEE International Symposium on Hardware Oriented Security and Trust (HOST)*, pp. 137–143, 2015.
6. K. Shamsi, M. Li, T. Meade, Z. Zhao, D. Z. Pan, and Y. Jin, "AppSAT: Approximately deobfuscating integrated circuits," in *2017 IEEE International Symposium on Hardware Oriented Security and Trust (HOST)*, pp. 95–100, 2017.
7. J. Rajendran, Y. Pino, O. Sinanoglu, and R. Karri, "Security analysis of logic obfuscation," in *Design Automation Conference*, pp. 83–89, 2012.
8. E. Nudelman, K. Leyton-Brown, H. H. Hoos, A. Devkar, and Y. Shoham, "Understanding random sat: Beyond the clauses-to-variables ratio," in *Principles and Practice of Constraint Programming – CP 2004* (M. Wallace, ed.), (Berlin, Heidelberg), pp. 438–452, Springer Berlin Heidelberg, 2004.
9. P. Chakraborty, J. Cruz, A. Alaql, and S. Bhunia, "SAIL: Analyzing structural artifacts of logic locking using machine learning," *IEEE Transactions on Information Forensics and Security*, pp. 1–1, 2021.
10. A. Taneem, D. K. Paul, and H. A. Jason, "Packing techniques for virtex-5 FPGAs," last accessed on Sep 25, 2023. Available at: https://janders.eecg.utoronto.ca/pdfs/trets_taneem.pdf.
11. M. Hansen, H. Yalcin, and J. Hayes, "Unveiling the ISCAS-85 benchmarks: a case study in reverse engineering," *IEEE Design & Test of Computers*, vol. 16, no. 3, pp. 72–80, 1999.
12. OpenCores, "DES cryptocore," last accessed on Sep 10, 2023. Available at: https://opencores.org/projects/basicdes.
13. OpenCores, "Basic RSA encryption engine," last accessed on Sep 10, 2023. Available at: https://opencores.org/projects/basicrsa.
14. OpenCores, "openGFX430," last accessed on Sep 10, 2023. Available at: https://opencores.org/projects/opengfx430.
15. OpenCores, "Classic 5-stage pipeline MIPS," last accessed on Sep 10, 2023. Available at: https://opencores.org/projects/mips32.
16. OpenCores, "JPEG decoder," last accessed on Sep 10, 2023. Available at: https://opencores.org/projects/jpeg_core.

17. OpenCores, "USB host core," last accessed on Sep 11, 2023. Available at: https://opencores.org/projects/usb_host_core.
18. OpenCores, "CORDIC core," last accessed on Sep 11, 2023. Available at: https://opencores.org/projects/cordic.
19. OpenCores, "Simple all digital FM receiver," last accessed on Sep 11, 2023. Available at: https://opencores.org/projects/all_digital_fm_receiver.
20. OpenCores, "2nd order sigma-delta DAC," last accessed on Sep 11, 2023. Available at: https://opencores.org/projects/sigma_delta_dac_dual_loop.
21. OpenCores, "openMSP430," last accessed on Sep 11, 2023. Available at: https://opencores.org/projects/openmsp430.
22. M. Imran, Z. U. Abideen, and S. Pagliarini, "An open-source library of large integer polynomial multipliers," in *2021 24th International Symposium on Design and Diagnostics of Electronic Circuits Systems (DDECS)*, pp. 145–150, 2021.
23. H. Hsing, "AES-128," last accessed on Jan 22, 2023. Available at: https://opencores.org/projects/tiny_aes.
24. H. Hsing, "SHAKE-256," last accessed on Mar 22, 2023. Available at: https://opencores.org/projects/sha3.
25. T. Zhu, "PID (proportional integral derivative) controller," last accessed on Dec 26, 2022. Available at: https://opencores.org/projects/pid_controller.
26. J. Carlos, "FPGA-based median filter," last accessed on Feb 19, 2023. Available at: https://opencores.org/projects/fpu100.
27. S. Joachim, "SHA-256," last accessed on Jan 20, 2023. Available at: https://github.com/secworks/sha256.
28. O. Kindgren and M. John, "OpenRISC 1200 implementation," last accessed on Feb 21, 2023. Available at: https://github.com/openrisc/or1200.
29. U. Embedded, "BiRiscV - 32-bit dual issue RISC-V CPU," last accessed on Feb 01, 2022. Available at: https://opencores.org/projects/biriscv.
30. J. Al-Eryani, "Floating-point unit (FPU) controller," last accessed on Feb 15, 2023. Available at: https://opencores.org/projects/fpu100.
31. FreeCores, "Infinite impulse response (IIR) filter," last accessed on Dec 25, 2021. Available at: https://github.com/freecores/all-pole_filters.
32. Z. U. Abideen, T. D. Perez, and S. Pagliarini, "From FPGAs to obfuscated eASICs: Design and security trade-offs," in *2021 Asian Hardware Oriented Security and Trust Symposium (AsianHOST)*, pp. 1–4, 2021.
33. Z. U. Abideen, T. D. Perez, M. Martins, and S. Pagliarini, "A security-aware and lut-based cad flow for the physical synthesis of hasics," *IEEE Transactions on Computer-Aided Design of Integrated Circuits and Systems*, vol. 42, no. 10, pp. 3157–3170, 2023.

Discussions and the Future of ReBO 12

12.1 A Fresh Look at ReBO as a Defense Technique

The ReBO techniques we presented encompass a wide range of methods, primarily falling into three distinct categories: LUT-based obfuscation (SRAM) [1–4], eFPGA redaction [5–7], and leveraging emerging technologies [8–11]. The LUT-based obfuscation methods focus on leveraging LUTs within SRAM to store bitstreams, providing a robust mechanism for securing the design against attacks. These techniques utilize the inherent flexibility of LUTs for obfuscation. They also manipulate the LUTs's placement during physical implementation, making it challenging for adversaries to decipher the original design. On the other hand, the eFPGA redaction category emphasizes the use of eFPGA macros that rely on SRAM for bitstream storage. This approach takes advantage of the reconfigurable nature of eFPGAs to obfuscate critical design components.

In addition to these established methods, the emerging approaches category explores innovative techniques that integrate LUTs with hybrid CMOS and emerging technologies. These next-generation obfuscation strategies aim to combine the best features of traditional CMOS technology with advancements in MRAM design, such as STT-MTJ and SOT-MTJ devices. By doing so, they target the robustness of PPA overheads during the obfuscation process, making it difficult for attackers to reverse engineer or tamper with the obfuscated hardware. Collectively, these techniques represent a comprehensive approach to ReBO, offering multiple layers of defense against an increasingly sophisticated threat landscape.

12.2 Review of Design Methods

In the context of ReBO, the primary goal is to identify and isolate specific parts of a design that can be mapped onto reconfigurable logic. This process requires a careful partitioning of the design into reconfigurable and non-reconfigurable logic, and this partitioning can occur

Z. U. Abideen and S. Pagliarini, *Reconfigurable Obfuscation Techniques for the IC Supply Chain*, Synthesis Lectures on Digital Circuits & Systems,
https://doi.org/10.1007/978-3-031-77509-3_12

at different abstraction levels, specifically at the HLS or RTL or gate-level, as discussed in Sect. 4.1.1.2.

When partitioning is performed at the HLS level, the design is divided early in the development process, where high-level functions and algorithms are targeted for mapping onto reconfigurable logic. An example of this approach can be found in [12], where researchers chose to partition their design at the HLS level, facilitating a more abstract and potentially faster path to hardware implementation. This method is particularly effective for applications like LUT-based obfuscation and eFPGA redaction, where higher-level design elements can be reconfigured to enhance security and flexibility.

When a hybrid approach is needed, with some parts of the design remaining static while others are reconfigurable, the most suitable method is to perform partitioning at the RTL or gate level. In the case of partitioning at the RTL, which is often referred to as dealing with "soft IP", it involves working with a more detailed design representation. This approach creates a balance between abstraction and control, enables accurate timing, efficient synthesis, and thorough verification.

Another example is given in [11], where the researchers partitioned their design on the RTL level and then mapped it to hardware [13]. Alternatively, the reconfigurable logic can be treated as a black box, and hard IP can be included in the final implementation design; a similar approach is also possible for gate-level partitioning.

12.3 Design Versus Security Trade-Offs

Balancing security and PPA overheads is crucial and is a function of the architecture and the designer's choices. The impact of changes in the architecture plays a vital role in the selection of reconfigurable parts. Concerning ReBO, almost all the researchers focus on security during the obfuscation phase [1, 6, 12, 14, 15]. Especially for the solution related to eFPGA redaction, they select a crucial but as small as possible part of the design so that it should not impact the PPA. Conversely, solutions leveraging emerging technologies prioritize the optimization of PPA alongside security [5, 16, 17]. However, early design decisions significantly influence the physical implementation of the design. The research in [18, 19] has developed a tool that designers can use to analyze the early trade-off between security and PPA overheads. This tool offers a means of understanding the impact of the reconfigurable part's size on security evaluation and its evolution over time.

12.4 Design Flow Automation and CAD Tools

Automating the selection of critical parts is challenging. An AI-based obfuscation engine trained on thousands of circuits may help to achieve this. However, if the circuit's architecture is novel or many circuit traces are different, it may pose a significant challenge.

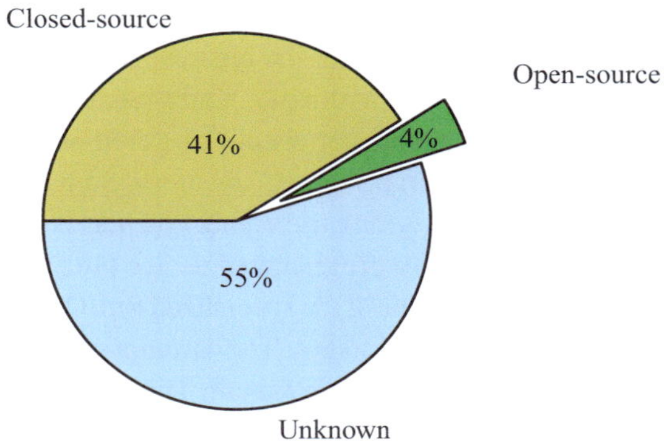

Fig. 12.1 Reconfigurable obfuscation techniques categorization based on developed/undeveloped CAD tools. The developed tools are split into open-source/closed-source

The majority of ReBO techniques require a specialized obfuscation tool that is made to be compatible with the traditional CAD flow. Often, researchers working on ReBO techniques develop custom tools [1–3, 6, 12, 14]. According to our findings, 41% of these techniques utilized custom tools for obfuscating the designs. These custom tools were closed-source but claimed compatibility with standard CAD processes as illustrated in Fig. 12.1). Interestingly, 59% of the techniques relied on standard CAD tools. It is noteworthy that a mere 4% of researchers opted to make their tools open-source. Regarding the remaining 55% of the techniques, the authors did not provide information about the availability of the tools, leaving it unclear whether they were open-source or closed-source.

12.5 Security and Storage of the Bitstream

The security of bitstreams is now essential in ReBO. These bitstreams contain the configuration information of an obfuscated IC. The bitstream may be vulnerable to attacks by adversaries seeking to exploit the design's IP or gain unauthorized access to sensitive information. The storage of bitstreams is critical and varies depending on the obfuscation class. Most of the studies related to ReBO did not discuss the long-term storage of their generated bitstreams [4–6, 12, 20–23]. It is likely that the researchers assumed the bitstream would be stored similarly to a standard FPGA configuration. LUT-based approach not only exploits SRAM for the storage [1–6, 12–14, 21, 24, 25], but some of the approaches also utilize FFs for the bitstream storage [18, 19]. SRAM is compact, highly dense, and low power, but it has to be utilized as a block in the physical implementation. On the other hand, FFs are standard cells from the commercial library, and they are less power efficient but more performance efficient.

In the context of emerging approaches, the bitstream is stored in a memory device that incorporates emerging technologies such as STT, SOT, MTJ, and MRAM. These storage devices are non-volatile, unlike SRAM. Therefore, the approaches based on these emerging technologies inherently facilitate long-term storage of bitstream [1–6, 12–14, 21, 24, 25]. As a result, the security of the bitstream becomes a concern. Notably, none of the research featured on this book has implemented bitstream encryption for security. Instead, it has been assumed that the bitstream would be stored externally and loaded at power-on or internally using a secure "tamper-proof" memory. Some researchers have suggested encrypting the bitstream using a symmetric cipher (such as AES) and eliminating the ability to read-back the bitstream from a programmed device [18, 19].

Furthermore, none of the specialized ReBO research discusses security during the deployment and distribution stages. The assumption is that all processes involving the activated IC occur within a secure environment. This involves ensuring that every aspect of the distribution, including loading the bitstream and activating the IC, takes place in a protected setting with security measures in place to prevent unauthorized access, tampering, or interception of data. A secure environment typically includes controlled facility access, encryption of the bitstream during transfer, authentication mechanisms, and adherence to stringent security protocols.

12.6 Lack of Silicon Validation

There is a significant shortage of frameworks to validate the security and functionality of obfuscated designs, as highlighted in [24], [?]. Most of the approaches for ReBO only synthesize the design and report the results. There is minimal effort towards silicon validation of the techniques as a proof of concept. For example, only techniques presented in [20, 24] were validated on silicon. Validating the techniques on silicon would provide motivation to advance reconfigurable-based obfuscation as a promising solution for protecting digital ICs.

12.6.1 Lack of Security Analysis

The ReBO approaches discussed in this book have mainly been assessed against SAT-based logical attacks [1–7, 12–14, 16, 17, 21, 24, 25]. Researchers typically conduct SAT attacks for at least 48 h to determine the resistance of a design. In some cases, they modify the LUT type, such as 2-input or 3-input LUT, and observe the execution time trend during the specified duration [26]. Two security metrics, execution time and variables-to-clauses ratio, were used in their analysis. While execution time is a common metric, more research is needed to accurately predict and interpret it [27]. The second metric, variables-to-clauses ratio, is useful, but it does not guarantee the security of designs meeting the ratio against SAT attacks, as evidenced in the hASIC analysis [19]. Instead, it can serve as a supplementary metric for security analysis.

On the other hand, there are a few new attacks that are developed to assess the security of ReBO [17, 19, 28, 29]. The attacks mentioned are specifically aimed at the ReBO approaches discussed in this book. They primarily exploit the structural traces to gain information and target the bitstream. However, there is a need for a more thorough security analysis to assess ReBO techniques against these specific attacks. Even for well-established attacks, there is still room for improvement in the analyses conducted so far.

12.7 Future Trends and Challenges

This book is an attempt to present the findings of various research papers in a clear manner. However, it is evident that the hardware security community lacks a unified benchmark suite and a common set of criteria for ReBo. Researchers often use benchmark suites popular within the testing community, but these might not be relevant to security. For example, the ISCAS'85 suite does not include crypto cores and other real applications crucial for evaluation in this field. Additionally, using circuits that better reflect current IC design practices, with IPs containing millions of gates and ICs housing billions of transistors, would greatly benefit the community.

The research community currently lacks standard criteria for evaluating the security of ReBO. While it is known that these techniques are resilient to advanced attacks, there is an urgent need for universally accepted criteria to assess their security. One of the generic metrics that we introduced is the ratio of gates to bitstream size as a simple but reasonable metric, but it is important to note that for eFPGA estimations, they need to be converted into the number of gates [30]. Several companies, including Achronix [31], Menta [32], and Quicklogic [33], offer eFPGA macros. Additionally, there is a noticeable trend in the ASIC design sector, with companies such as AMD [34] and Intel [35] acquiring FPGA technology. This strategic move allows these companies to benefit from both ASIC and FPGA domains. This convergence of the two technologies could lead to reconfigurable logic becoming an integral aspect of commercial production for security purposes. However, it is worth noting that the driving force behind ASIC-FPGA hybrid solutions today is design metrics and goals rather than security.

It is also worth discussing the threat models proposed so far. The authors of [19] have established a robust threat model. Defining the capabilities of an attacker remains a complex task, requiring an understanding of their motivations, technical skills, and resource availability. Underestimating the attacker may result in ineffective defense strategies, whereas overestimating them could lead to unnecessary PPA overheads due to convoluted defense strategies. This challenge extends to ReBO and any other obfuscation-promoting approaches.

Another consideration in terms of attack is whether an attacker can leverage a partially recovered netlist. For instance, in a design that employs multiple instances of the same block, correctly recovering one block may enable the attacker to recover all other instances through a visual inspection of their structure [36]. This line of thinking is also applicable to datapaths and certain cryptographic structures that exhibit regularity. Consequently, a functional analysis of the recovered netlist can be combined with existing attacks to enhance correctly guessed connections.

The information presented in Fig. 12.1 indicates that creating a custom tool is an essential aspect of the obfuscation process, with 45% of researchers developing their own tools, and 41% of these tools being closed source. This closed-source practice presents a barrier to the research community, and there should be an effort to encourage the open-sourcing of these tools for academic use. Future research should investigate how these tools, in combination with reverse engineering and other mentioned methods, can be leveraged to compromise the security of ReBO. LL is a cautionary tale; LL has become increasingly vulnerable to SAT attacks, despite various attempted defenses. Additionally, ReBO techniques may eventually surpass LL due to their inherent security against SAT attacks. There is an anticipation that eFPGA redaction offers better security than LL, and LL is almost "dead" due to the cat and mouse game. Researchers in this domain will likely benefit from open-source macro generators for eFPGAs, as only commercial solutions currently exist. A significant amount of research publications is expected to emerge in this area in the coming years [17, 28, 29].

References

1. A. Baumgarten, A. Tyagi, and J. Zambreno, "Preventing IC piracy using reconfigurable logic barriers," *IEEE Design Test of Computers*, vol. 27, no. 1, pp. 66–75, 2010.
2. H. Mardani Kamali, K. Zamiri Azar, K. Gaj, H. Homayoun, and A. Sasan, "LUT-lock: A novel LUT-based logic obfuscation for FPGA-bitstream and ASIC-hardware protection," in *2018 IEEE Computer Society Annual Symposium on VLSI (ISVLSI)*, pp. 405–410, 2018.
3. S. D. Chowdhury, G. Zhang, Y. Hu, and P. Nuzzo, "Enhancing SAT-attack resiliency and cost-effectiveness of reconfigurable-logic-based circuit obfuscation," in *2021 IEEE International Symposium on Circuits and Systems (ISCAS)*, pp. 1–5, IEEE, 2021.
4. J. Bhandari, A. K. Thalakkattu Moosa, B. Tan, C. Pilato, G. Gore, X. Tang, S. Temple, P.-E. Gaillardon, and R. Karri, "Exploring eFPGA-based redaction for IP protection," in *2021 IEEE/ACM International Conference On Computer Aided Design (ICCAD)*, pp. 1–9, 2021.
5. C. M. Tomajoli, L. Collini, J. Bhandari, A. K. T. Moosa, B. Tan, X. Tang, P.-E. Gaillardon, R. Karri, and C. Pilato, "ALICE: An automatic design flow for eFPGA redaction," in *Proceedings of the 59th ACM/IEEE Design Automation Conference*, p. 781–786, 2022.
6. P. Mohan, O. Atli, J. Sweeney, O. Kibar, L. Pileggi, and K. Mai, "Hardware redaction via designer-directed fine-grained eFPGA insertion," in *2021 Design, Automation & Test in Europe Conference & Exhibition (DATE)*, pp. 1186–1191, IEEE, 2021.
7. J. Bhandari, A. K. T. Moosa, B. Tan, C. Pilato, G. Gore, X. Tang, S. Temple, P.-E. Gaillardon, and R. Karri, "Not all fabrics are created equal: Exploring efpga parameters for ip redaction," *IEEE Trans. Very Large Scale Integr. Syst.*, vol. 31, p. 1459–1471, oct 2023.
8. T. Winograd, H. Salmani, H. Mahmoodi, K. Gaj, and H. Homayoun, "Hybrid STT-CMOS designs for reverse-engineering prevention," in *Proceedings of the 53rd Annual Design Automation Conference*, pp. 1–6, 2016.
9. J. Yang, X. Wang, Q. Zhou, Z. Wang, H. Li, Y. Chen, and W. Zhao, "Exploiting spin-orbit torque devices as reconfigurable logic for circuit obfuscation," *IEEE Transactions on Computer-Aided Design of Integrated Circuits and Systems*, vol. 38, no. 1, pp. 57–69, 2018.

10. G. Kolhe, S. Salehi, T. D. Sheaves, H. Homayoun, S. Rafatirad, M. P. Sai, and A. Sasan, "Securing hardware via dynamic obfuscation utilizing reconfigurable interconnect and logic blocks," in *2021 58th ACM/IEEE Design Automation Conference (DAC)*, pp. 229–234, IEEE, 2021.
11. M. M. Shihab, B. Ramanidharan, S. S. Tellakula, G. Rajavendra Reddy, J. Tian, C. Sechen, and Y. Makris, "ATTEST: Application-agnostic testing of a novel transistor-level programmable fabric," in *2020 IEEE 38th VLSI Test Symposium (VTS)*, pp. 1–6, 2020.
12. J. Chen, M. Zaman, Y. Makris, R. D. S. Blanton, S. Mitra, and B. C. Schafer, "DECOY: Deflection-Driven HLS-Based Computation Partitioning for Obfuscating Intellectual PropertY," in *Proceedings of the 57th ACM/EDAC/IEEE Design Automation Conference*, DAC '20, IEEE Press, 2020.
13. B. Liu and B. Wang, "Embedded reconfigurable logic for ASIC design obfuscation against supply chain attacks," in *2014 Design, Automation Test in Europe Conference Exhibition (DATE)*, pp. 1–6, 2014.
14. B. Hu, T. Jingxiang, S. Mustafa, R. R. Gaurav, S. William, M. Yiorgos, C. S. Benjamin, and S. Carl, "Functional obfuscation of hardware accelerators through selective partial design extraction onto an embedded FPGA," in *Proceedings of the 2019 Great Lakes Symposium on VLSI*, p. 171–176, 2019.
15. Z. U. Abideen, S. Gokulanathan, M. J. Aljafar, and S. Pagliarini, "An overview of FPGA-inspired obfuscation techniques," *Association for Computing Machinery*, vol. 56, no. 12, December 2024
16. P. Mohan, O. Atli, O. Kibar, M. Zackriya, L. Pileggi, and K. Mai, "Top-down physical design of soft embedded FPGA fabrics," in *The 2021 ACM/SIGDA International Symposium on Field-Programmable Gate Arrays*, p. 1–10, 2021.
17. A. Rezaei, R. Afsharmazayejani, and J. Maynard, "Evaluating the security of eFPGA-based redaction algorithms," ICCAD '22, (New York, NY, USA), Association for Computing Machinery, 2022.
18. Z. U. Abideen, T. D. Perez, and S. Pagliarini, "From FPGAs to obfuscated eASICs: Design and security trade-offs," in *2021 Asian Hardware Oriented Security and Trust Symposium (AsianHOST)*, pp. 1–4, 2021.
19. Z. U. Abideen, T. D. Perez, M. Martins, and S. Pagliarini, "A security-aware and lut-based cad flow for the physical synthesis of hasics," *IEEE Transactions on Computer-Aided Design of Integrated Circuits and Systems*, vol. 42, no. 10, pp. 3157–3170, 2023.
20. M. M. Shihab, J. Tian, G. R. Reddy, B. Hu, W. Swartz, B. Carrion Schaefer, C. Sechen, and Y. Makris, "Design obfuscation through selective post-fabrication transistor-level programming," in *2019 Design, Automation & Test in Europe Conference & Exhibition (DATE)*, pp. 528–533, 2019.
21. J. Chen and B. C. Schafer, "Area efficient functional locking through coarse grained runtime reconfigurable architectures," in *Proceedings of the 26th Asia and South Pacific Design Automation Conference*, pp. 542–547, 2021.
22. S. Patnaik, N. Rangarajan, J. Knechtel, O. Sinanoglu, and S. Rakheja, "Advancing hardware security using polymorphic and stochastic spin-hall effect devices," in *2018 Design, Automation & Test in Europe Conference & Exhibition (DATE)*, pp. 97–102, IEEE, 2018.
23. N. Rangarajan, S. Patnaik, J. Knechtel, R. Karri, O. Sinanoglu, and S. Rakheja, "Opening the doors to dynamic camouflaging: Harnessing the power of polymorphic devices," *IEEE Transactions on Emerging Topics in Computing*, 2020.
24. G. Kolhe, T. Sheaves, K. I. Gubbi, T. Kadale, S. Rafatirad, S. M. PD, A. Sasan, H. Mahmoodi, and H. Homayoun, "Silicon validation of LUT-based logic-locked IP cores," in *Proceedings of the 59th ACM/IEEE Design Automation Conference*, pp. 1189–1194, 2022.
25. C. Sathe, Y. Makris, and B. C. Schafer, "Investigating the effect of different eFPGAs fabrics on logic locking through HW redaction," in *2022 IEEE 15th Dallas Circuit And System Conference (DCAS)*, pp. 1–6, IEEE, 2022.

26. G. Kolhe, H. M. Kamali, M. Naicker, T. D. Sheaves, H. Mahmoodi, P. D. Sai Manoj, H. Homayoun, S. Rafatirad, and A. Sasan, "Security and complexity analysis of LUT-based obfuscation: From blueprint to reality," in *2019 IEEE/ACM International Conference on Computer-Aided Design (ICCAD)*, pp. 1–8, 2019.
27. B. Ahmed, S. Rahman, K. Z. Azar, F. Farahmandi, F. Rahman, and M. Tehranipoor, "Seemless: Security evaluation of logic locking using machine learning oriented estimation," in *Proceedings of the Great Lakes Symposium on VLSI 2024*, GLSVLSI '24, p. 489–494, Association for Computing Machinery, 2024.
28. P. Chowdhury, C. Sathe, and B. Carrion Schaefer, "Predictive model attack for embedded FPGA logic locking," in *Proceedings of the ACM/IEEE International Symposium on Low Power Electronics and Design*, pp. 1–6, 2022.
29. Z. Han, M. Shayan, A. Dixit, M. Shihab, Y. Makris, and J. J. Rajendran, "FuncTeller: How well does eFPGA hide functionality?," in *32nd USENIX Security Symposium (USENIX Security 23)*, pp. 5809–5826, 2023.
30. Z. U. Abideen, S. Gokulanathan, M. J. Aljafar, and S. Pagliarini, "An overview of FPGA-inspired obfuscation techniques," *ACM Comput. Surv.*, jul 2024. Just Accepted.
31. Achronix Corp., "Speedcore embedded FPGA IP," last accessed on Apr 22, 2023. Available at: https://www.achronix.com/product/speedcore.
32. Menta, "Embedded FPGA IP," last accessed on Apr 2, 2022. Available at: https://www.menta-efpga.com/.
33. QuickLogic Corp., "efpga ip 2.0 – enabling mass customization with fpga technology," last accessed on Jan19, 2023. Available at: https://www.quicklogic.com/products/efpga/efpga-ip2/.
34. Xilinx, Inc., "AMD acquires xilinx," last accessed on Mar 29, 2023. Available at: https://www.amd.com/en/corporate/xilinx-acquisition.
35. Intel Corp., "Intel acquisition of altera," last accessed on Apr 15, 2023. Available at: https://newsroom.intel.com/press-kits/intel-acquisition-of-altera/.
36. G. Basiashvili, Z. U. Abideen, and S. Pagliarini, "Obfuscating the hierarchy of a digital ip," in *Embedded Computer Systems: Architectures, Modeling, and Simulation* (A. Orailoglu, M. Reichenbach, and M. Jung, eds.), (Cham), pp. 303–314, Springer International Publishing, 2022.

Securing the Bitstream of hASIC

A

Security Starts Where Users Engage

A.0.1 Protecting the End-User's Security

The security of bitstreams has become crucial to protecting ICs deployed in various products. The bitstream, which contains the configuration information of an obfuscated IC, can be vulnerable to attacks by adversaries seeking to exploit the design's IP or gain unauthorized access to sensitive information. To counter these threats, cryptographic techniques are employed to safeguard the bitstream's security. One of the primary concerns regarding bitstream security is the potential for adversaries to leverage an oracle and reverse engineer the bitstream. In this context, encryption involves the generation of a robust key for encryption. To establish strong encryption, a robust key generation mechanism is essential.

A.0.2 Introduction to PUF as a Security Primitive

PUFs are hardware-based security components that exploit inherent physical variations within integrated circuits. Since PUF responses are unique to each IC, they prevent the cloning or tampering of secret key. These variations are unique to each IC, providing a reliable and unclonable identity. PUF leverages these unique characteristics to generate cryptographic keys, serving as a *foundation for secure key generation* [1, 2]. PUFs exploit process variation (e.g., gate oxide thickness, size, and threshold voltage) that occurs naturally during the fabrication of ICs. Although the circuits are fabricated with identical layouts, every transistor presents slightly random electric properties that generate a unique identity [3]. By utilizing PUFs as a root of trust for key generation, a strong foundation is established for the encryption of bitstreams [4]. This enhances the overall security of the encryption process

Z. U. Abideen and S. Pagliarini, *Reconfigurable Obfuscation Techniques for the IC Supply Chain*, Synthesis Lectures on Digital Circuits & Systems,
https://doi.org/10.1007/978-3-031-77509-3

and ensures that only authorized entities with access to the correct PUF response can decrypt and access the bitstream.

PUFs generate a set of challenge-response pairs (CRPs) that can be classified into two categories: extensive PUFs and confined PUFs [5]. Extensive PUFs provide exponential CRPs, whereas confined PUFs offer only one or a few CRPs. Different types of PUFs can be classified into various groups, including ring oscillator-based PUF (RO-PUF) [6], arbiter PUF [7], DRAM PUF [8], and SRAM-based PUF [9–17]. Examples of confined PUFs include SRAM-based PUFs, such as those mentioned in [9–17]. These types of PUFs are commonly used for storing unique identifiers or long-term secret keys [18]. On the other hand, an extensive PUF can accept multiple challenges and generate a 1-bit response to an n-bit challenge, inducing a random n-variable Boolean function from a computational perspective. Examples of extensive PUFs include RO-PUFs [6] and arbiter PUFs [7]. Extensive PUFs are typically utilized for challenge-response authentication [4].

The digital signature of an SRAM-based PUF is the raw entropy of the SRAM array that is converted into digital bits. SRAM-based PUFs offer a combination of simplicity, low cost, high reliability, and scalability [19]. Additionally, SRAM-based PUFs rely on standard SRAM IP that is commonly available to designers, and the same memory macro utilized for storage can also serve as a PUF. There is no need to customize the memory macro, and the internal architecture of 6T-SRAM cell is illustrated in Fig. A.1.

In a standard 6T SRAM, each bitcell comprises six transistors, including two cross-coupled CMOS inverters and two access transistors, as given in Fig. A.1. The control signal to access the SRAM bitcell is line WL. During read and write operations, bit lines BL and $\overline{BL}$ carry data. Two signals, namely Q and $\overline{Q}$ are internal signals and one of them becomes output when driving the bit lines BL and $\overline{BL}$. The four transistors placed symmetrically in Fig. A.1 form bistable inverters. These inverters are symmetrically designed to match size, but there

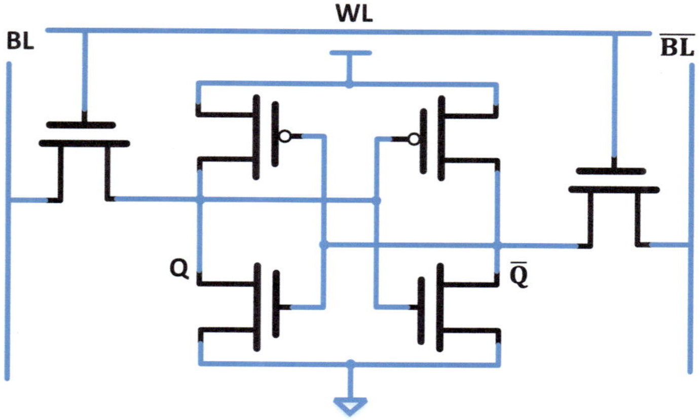

Fig. A.1 The internal architecture of 6T-SRAM bitcell, adapted from [20]

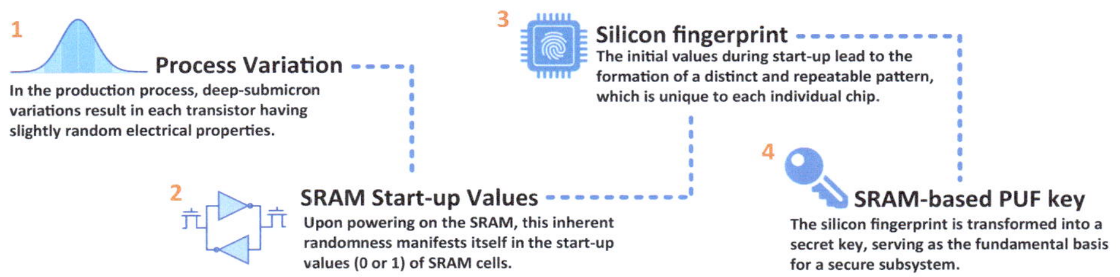

Fig. A.2 The procedure to harvest unique signature from SRAM-PUF [22]

may be mismatches due to random variations during fabrication. These mismatches can be used by SRAM-based PUFs, which take advantage of biases in each SRAM bitcell towards a logic '0' or logic '1' when powered up. Since CMOS devices have different physical parameters during fabrication, such as doping levels and transistor oxide thickness, these variations can affect the power-up state of associated bitcells in an SRAM. Some bitcells may strongly prefer a logic '0' or logic '1' state upon power-up, while others are neutral and power up randomly due to system noise. The cells that strongly prefer a logic '0' or logic '1' state are more useful for PUF response. It has been shown in [19] that not all platforms can function as PUF. By examining the power-up state of an SRAM, a unique identifier can be created since the process variations and preferences are truly random and dependent on physical anomalies. The process of generating a key starts by extracting the PUF response, which is a distinct and unpredictable value derived from the physical characteristics of the IC, as demonstrated in Fig. A.2. This response is utilized to generate secret keys that are specific to the cryptographic algorithms. The uniqueness and randomness of PUF responses contribute to the strength and security of the generated keys.

A.0.2.1 Robustness of SRAM-Based PUF

The output of SRAM-based PUF must stay consistent even when voltage and temperature conditions change [23]. SRAM bitcells can be split into two groups based on their power-up value, which depends entirely on the strength between the two cross-coupled inverters within the SRAM bitcell [22]. Below are the two types of bitcells for SRAM-based PUF:

- Neutral bitcell: The strength difference between two cross-coupled inverters in an SRAM bitcell is minimal. The power-up value of a bitcell may rely on measurement noise and random logic '0' and logic '1' states, as illustrated in Fig. A.3.
- Skewed bitcell: While powering up an SRAM bitcell, the strength difference between two cross-coupled inverters is crucial in determining whether a logic '0' or logic '1' is produced. However, some cells may have little process variation, resulting in a weak logic '0' or logic '1'. The threshold voltage (V_{th}) of transistors is the most significant factor in determining the start-up value of an SRAM bitcell, as reported in [25]. As the CMOS

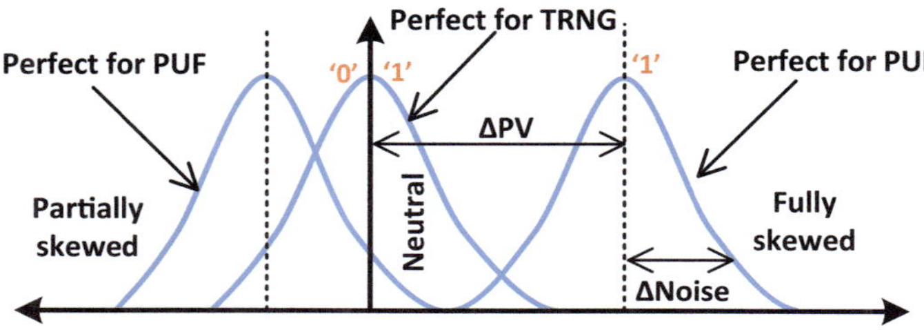

Fig. A.3 Characteristic of SRAM bitcells on power-up state, adapted from [23]

technology node shrinks, intra-die variability increases, leading to process variability. This variability in fabrication processes results in an increased difference in strength between two cross-coupled inverters, which improves the quality of SRAM-based PUF. These partially skewed cells are suitable for true random number generator (TRNG) and PUF applications in identification where errors can be tolerated. However, power-up values of these bitcells can be impacted by measurement noise, temperature and voltage fluctuations, and aging. On the other hand, a strong mismatch between cross-coupled inverters in an SRAM bitcell may produce a strong logic '0' or logic ''1', as depicted in Fig. A.3.

A.0.2.2 Evaluation Metrics for SRAM-Based PUF

The quality of an SRAM-based PUF can be defined by its reliability, entropy, uniqueness, and randomness. The *reliability* metric is a key characteristic that defines the ability of a PUF to consistently reproduce its output response, independent of temperature variations and fluctuations in operating voltage. The SRAM-based PUF must generate the same response at all operating conditions in every power-up cycle during its entire lifetime. The reliability of its PUF can be assessed by evaluating its with-in class hamming distance (WCHD), which is the fractional hamming distance between measurements taken at various conditions during reconstruction and a reference measurement taken at enrollment. Another reliability-related metric is within class sequential hamming distance (WCSHD), the Hamming distance between the consecutive responses of the same SRAM-based PUF, defined in [26]. The WCHD can be calculated by Eq. A.1. Where R_j denotes the reference n-bit response extracted at the normal operating condition (i.e., room temperature and the normal supply voltage) for a chip j. $\overline{R}_{j,s}$ indicates the response at normal operating conditions s, and x is the total number of responses for the PUF.

$$\text{WCHD} = \frac{1}{x} \sum_{S=1}^{x} \frac{HD\left(R_j, \overline{R}_{j,s}\right)}{n} \times 100\% \tag{A.1}$$

For *entropy*, one important parameter is a bias pattern linked with the PUF's characteristics. Understanding the origin of *bias pattern* is crucial for improving the reliability and security of SRAM-based PUFs. SRAM-based PUFs can be attributed to fabrication variations, temperature changes, or other factors. This requires a comprehensive analysis of their response data. In particular, the response data shows that a set of bitcells within the output response exhibit a consistent bias towards a logic '0' or logic '1', which cannot directly be assessed by fractional hamming weight. This is defined by the percentage of ones in the raw output response. This recurrent phenomenon is usually referred to as the bias pattern. SRAM-based PUFs leverage *entropy* from process variations to build various kinds of unique fingerprints for each identically fabricated chip. Entropy in SRAM-based PUF is derived by inserting the masked hamming weight (MHW) into min-entropy formula and MHW is calculated through the percentage of ones in the output response after an XOR operation with the bias pattern [27]. The mathematical form for the MHW and min-entropy is presented in Eqs. A.2 and A.3.

$$MHW = \frac{1}{n}\sum_{i=1}^{n} R_i \oplus m_i \tag{A.2}$$

$$H_{\min} = -\log_2(\max(MHW, 1 - MHW)) \tag{A.3}$$

The value of the SRAM-based PUF's response (R) on bitcell i is denoted as R_i. This value is derived with the bias pattern (m) value on bit position i. Both the SRAM-based PUF response and the bias pattern have a length of n bits. The metric of *uniqueness* is a determination of the capability of an SRAM-based PUF to produce unique responses across multiple chips. Different SRAM-based PUFs must generate different responses to a given challenge to separate one from another. Different chips may produce nearly identical PUF responses due to systematic variations, and this is measured by bias. The concept of uniqueness can be quantitatively expressed as a between-class hamming distance (BCHD) in Eq. A.4. Where R_i and R_j represent the responses produced by two different chips (i, j) upon the application of the same challenge. n represents the length of the response, and N signifies the total number of chips. In an ideal scenario, the uniqueness, quantified by the BCHD, should be equal to 50%.

$$\text{BCHD} = \frac{2}{N\,(N+1)}\sum_{i=1}^{N-1}\sum_{j=i+1}^{N}\frac{HD\left(R_i, R_j\right)}{n} \times 100\% \tag{A.4}$$

The metric of *randomness* defines the random response generated by a PUF regarding the probability distribution of the logic '0' and logic '1' states. Ideally, the randomness should be balanced at 50%, meaning the probability of obtaining a logic '0' response must be equal to the probability of obtaining a logic '1' response. Randomness also helps to determine whether a PUF is biased or not. For an unbiased SRAM-based PUF, changing one bit in a challenge or address of SRAM should alter approximately half of the response bits.

References

1. Abideen, Zain Ul and Wang, Rui and Perez, Tiago Diadami and Schrijen, Geert-Jan and Pagliarini, Samuel, IEEE Design & Test," *Impact of Orientation on the Bias of SRAM-Based PUFs*, vol. 41, no. 3, 2024.
2. Aljafar, Muayad J. and Ul Abideen, Zain and Peetermans, Adriaan and Gierlichs, Benedikt and Pagliarini, Samuel, "IEEE Embedded Systems Letters", *SCALLER: Standard Cell Assembled and Local Layout Effect-Based Ring Oscillators*, vol. 16, no. 4, 2024.
3. B. Gassend, D. Clarke, M. van Dijk, and S. Devadas, "Silicon physical random functions," in *Proceedings of the 9th ACM Conference on Computer and Communications Security*, CCS '02, p. 148-160, Association for Computing Machinery, 2002.
4. G. E. Suh and S. Devadas, "Physical unclonable functions for device authentication and secret key generation," in *2007 44th ACM/IEEE Design Automation Conference*, pp. 9–14, 2007.
5. International Organization for Standardization, "ISO/IEC 20897-1:2020 information security, cybersecurity and privacy protection - physically unclonable functions - part 1: Security requirements," last accessed on Jun 30, 2020. Available at: https://www.iso.org/standard/76353.html.
6. S. S. Mansouri and E. Dubrova, "Ring oscillator physical unclonable function with multi level supply voltages," in *2012 IEEE 30th International Conference on Computer Design (ICCD)*, pp. 520–521, 2012.
7. K. Fruhashi, M. Shiozaki, A. Fukushima, T. Murayama, and T. Fujino, "The arbiter-PUF with high uniqueness utilizing novel arbiter circuit with delay-time measurement," in *2011 IEEE International Symposium of Circuits and Systems (ISCAS)*, pp. 2325–2328, 2011.
8. J. Miskelly and M. O'Neill, "Fast DRAM PUFs on commodity devices," *IEEE Transactions on Computer-Aided Design of Integrated Circuits and Systems*, vol. 39, no. 11, pp. 3566–3576, 2020.
9. S. Zhang, B. Gao, D. Wu, H. Wu, and H. Qian, "Evaluation and optimization of physical unclonable function (PUF) based on the variability of FinFET SRAM," in *2017 International Conference on Electron Devices and Solid-State Circuits (EDSSC)*, pp. 1–2, 2017.
10. K.-H. Chuang, E. Bury, R. Degraeve, B. Kaczer, D. Linten, and I. Verbauwhede, "A physically unclonable function using soft oxide breakdown featuring 0% native BER and 51.8 fj/bit in 40-nm CMOS," *IEEE Journal of Solid-State Circuits*, vol. 54, no. 10, pp. 2765–2776, 2019.
11. R. Maes, V. Rozic, I. Verbauwhede, P. Koeberl, E. van der Sluis, and V. van der Leest, "Experimental evaluation of physically unclonable functions in 65 nm CMOS," in *2012 Proceedings of the ESSCIRC (ESSCIRC)*, pp. 486–489, 2012.
12. S. Baek, G.-H. Yu, J. Kim, C. T. Ngo, J. K. Eshraghian, and J.-P. Hong, "A reconfigurable SRAM based CMOS PUF with challenge to response pairs," *IEEE Access*, vol. 9, pp. 79947–79960, 2021.
13. Y. Shifman, A. Miller, Y. Weizmann, and J. Shor, "A 2 bit/cell tilting sram-based PUF with a BER of 3.1e-10 and an energy of 21 fj/bit in 65nm," *IEEE Open Journal of Circuits and Systems*, vol. 1, pp. 205–217, 2020.
14. A. B. Alvarez, W. Zhao, and M. Alioto, "Static physically unclonable functions for secure chip identification with 1.9-5.8% native bit instability at 0.6-1 v and 15 fj/bit in 65 nm," *IEEE Journal of Solid-State Circuits*, vol. 51, no. 3, pp. 763–775, 2016.
15. G.-J. Schrijen and V. van der Leest, "Comparative analysis of SRAM memories used as PUF primitives," in *2012 Design, Automation & Test in Europe Conference & Exhibition (DATE)*, pp. 1319–1324, 2012.
16. G. Selimis, M. Konijnenburg, M. Ashouei, J. Huisken, H. de Groot, V. van der Leest, G.-J. Schrijen, M. van Hulst, and P. Tuyls, "Evaluation of 90nm 6t-sram as physical unclonable function for secure key generation in wireless sensor nodes," in *2011 IEEE International Symposium of Circuits and Systems (ISCAS)*, pp. 567–570, 2011.

17. R. Wang, G. Selimis, R. Maes, and S. Goossens, "Long-term continuous assessment of SRAM PUF and source of random numbers," in *2020 Design, Automation & Test in Europe Conference & Exhibition (DATE)*, pp. 7–12, 2020.
18. A. Shamsoshoara, A. Korenda, F. Afghah, and S. Zeadally, "A survey on physical unclonable function (PUF)-based security solutions for internet of things," *Computer Networks*, vol. 183, p. 107593, 2020.
19. Intrisic ID, "SRAM PUF: The secure silicon fingerprint," last accessed on Sep 11, 2023. Available at: https://www.intrinsic-id.com/wp-content/uploads/2023/03/2023-03-09-White-Paper-SRAM-PUF-The-Secure-Silicon-Fingerprint.pdf.
20. G. Srinivasan, P. Wijesinghe, S. S. Sarwar, A. Jaiswal, and K. Roy, "Significance driven hybrid 8T-6T SRAM for energy-efficient synaptic storage in artificial neural networks," in *2016 Design, Automation & Test in Europe Conference & Exhibition (DATE)*, pp. 151–156, 2016.
21. A. Van Herrewege, A. Schaller, S. Katzenbeisser, and I. Verbauwhede, "DEMO: Inherent PUFs and secure PRNGs on commercial off-the-shelf microcontrollers," in *Proceedings of the 2013 ACM SIGSAC Conference on Computer & Communications Security*, CCS '13, (New York, NY, USA), p. 1333-1336, Association for Computing Machinery, 2013.
22. Intrinsic ID, "SRAM PUF technology," last accessed on Jul 01, 2022. Available at: https://www.intrinsic-id.com/sram-puf/.
23. M. T. Rahman, A. Hosey, Z. Guo, J. Carroll, D. Forte, and M. Tehranipoor, "Systematic correlation and cell neighborhood analysis of SRAM PUF for robust and unique key generation," *Journal of Hardware and Systems Security*, vol. 1, pp. 137–155, Jun 2017.
24. D. E. Holcomb, W. P. Burleson, and K. Fu, "Power-up SRAM state as an identifying fingerprint and source of true random numbers," *IEEE Transactions on Computers*, vol. 58, no. 9, pp. 1198–1210, 2009.
25. M. Cortez, S. Hamdioui, V. van der Leest, R. Maes, and G.-J. Schrijen, "Adapting voltage ramp-up time for temperature noise reduction on memory-based PUFs," in *2013 IEEE International Symposium on Hardware-Oriented Security and Trust (HOST)*, pp. 35–40, 2013.
26. S. Elgendy and E. Y. Tawfik, "Impact of physical design on PUF behavior: A statistical study," in *2021 IEEE International Symposium on Circuits and Systems (ISCAS)*, pp. 1–5, 2021.
27. National Institute of Standards and Technology (NIST), "Recommendation for the entropy sources used for random bit generation," last accessed on Oct 20, 2018. Available at: https://nvlpubs.nist.gov/nistpubs/SpecialPublications/NIST.SP.800-90B.pdf.

Securing the Bitstream of hASIC

A

Security Starts Where Users Engage

A.1 Encrypting the Bitstream of hASIC

The security of the bitstream is a critical aspect of obfuscation, as it faces threats from the end-user, who may attempt to tamper with or expose the protected design. To enhance design security, utilizing a cryptocore to encrypt the bitstream is a viable option. hASIC encryption scheme employs the AES algorithm, ensuring that the bitstream is protected from unauthorized access. The level of security provided by the encrypted bitstream is highly reliable, as it cannot be copied or reverse engineered. The encryption scheme uses AES-256. NIST states that approximately are approximately 1.1×10^{77} possible combinations for a 256-bit key [1]. Symmetric encryption algorithms, like AES, use the same key for encryption and decryption. The safety of the data is directly related to the confidentiality of the key. The security of bitstream encryption relies on the confidentiality of the key.

The process of generating a secret key, encrypting, and decrypting a bitstream is illustrated in Fig. A.1. The encryption chip contains an SRAM-based PUF, error correction logic, control logic, and an AES encryption block for encrypting the bitstream. The SRAM-based PUF generates a unique secret key on the power-up state of bitcells. However, each time the SRAM starts up, a slightly different pattern may emerge, creating a noise component dependent on temperature, voltage ramp, and operating conditions. Despite this noise, it is possible to reconstruct a reliable key every time the SRAM is powered, thanks to error correction, such as "helper data algorithms" [2]. The hASIC bitstream is encrypted with a secret key. The output of the encryption chips is an encrypted bitstream and the secret key. This entire process takes place in a trusted environment.

Z. U. Abideen and S. Pagliarini, *Reconfigurable Obfuscation Techniques for the IC Supply Chain*, Synthesis Lectures on Digital Circuits & Systems,
https://doi.org/10.1007/978-3-031-77509-3

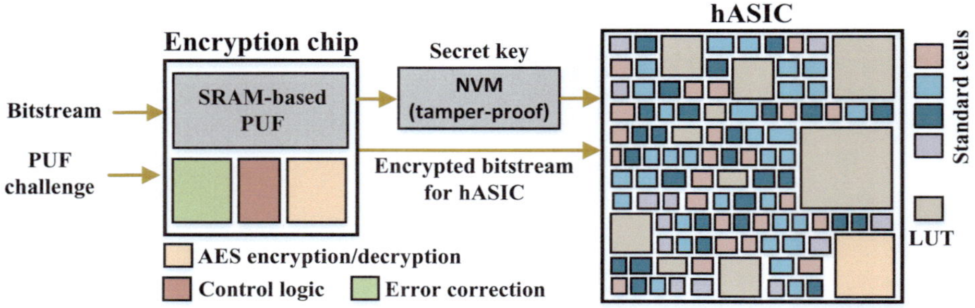

Fig. A.1 Encryption/decryption scheme of hASIC with SRAM-based PUF

At a trusted facility, the hASIC is loaded with the secret key from a tamper-proof memory before being fed the bitstream. During configuration, hASIC performs the opposite process by decrypting the incoming bitstream, as illustrated in Fig. A.1. The encryption logic employed by hASIC uses a 256-bit encryption key. It is essential to note that the AES decryption logic in hASIC is solely dedicated to bitstream decryption and cannot be utilized for any other purpose. hASIC comprises static logic that takes the form of standard cells and reconfigurable logic in the form of LUTs. This way, the AES decryption logic is also integrated into the hASIC as a block, as depicted in Fig. A.1. If the origin of the encryption key can be trusted and the keys are extracted securely in hardware, they form the so-called "root of trust" of the device. As the security of a bitstream depends solely on the secret key, the robustness of SRAM-based PUFs is crucial in this analysis. The SRAM-based PUF should generally satisfy certain characteristics, as described in the previous appendix.

A.1.1 Internal Architecture of SRAM

This section explains the internal architecture of SRAM before presenting the design of SRAM-based PUF and its evaluation. SRAM relies on the bitcell, consisting of two CMOS inverters connected in a positive feedback loop, to form a bistable storage element. The initial state of each bitcell is determined by the process variation that occurs during the IC's manufacturing process. The stability of each bit is dependent on the degree of threshold voltage mismatch between the local devices. The typical 6T-SRAM cell has a preferred state due to stochastic variations in the threshold voltages of its transistors. The randomness in the initial values of 6T-SRAM results in an unpredictable yet repeatable pattern of zeros and ones that are unique to each device.

Figure A.2 clearly illustrates the high-level information of memory architecture. During the placement phase of an ASIC, the designer can rotate memories. Figure A.2 illustrates some possible memory rotations. Two types of memories from a major foundry were considered in this study: high-speed and low-density, and low-speed and high-density. The

Fig. A.2 The simplified architecture and orientations of memory [3]

high-speed memory uses standard threshold voltage for both the periphery and bitcells. In contrast, low-speed memory employs mixed threshold voltage for the periphery and high threshold voltage for bitcells. The SRAMs are arranged in an array of memory locations, where each memory access involves reading or writing all the bits in a single location. SRAM macros can be organized in various ways depending on the user's specification for the desired addresses and datawidth. The memory compiler automatically makes these decisions.

The detailed architecture of a single port low-speed memory is shown in Fig. A.3. Often, commercial SRAM compilers generate memories with half of the bits on the right and the other half on the left. The control circuitry is located in the center. This arrangement is identical for high-speed and high-density variants. To create large bitcell arrays, a memory matrix (M) of size $j \times k$ is replicated multiple times to form a larger matrix of size $M \times C$. The memory compiler determines the aspect ratio of the memory and the number of bitcells

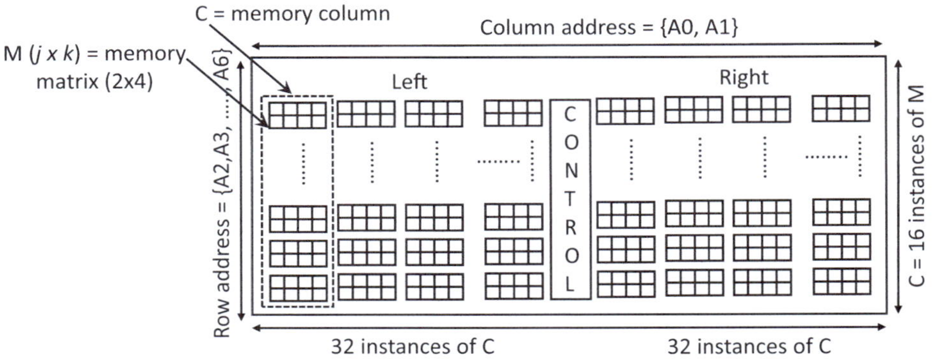

Fig. A.3 Architecture of low-speed memory with 64-bit datawidth and 128 location depth

that need to be Muxed together by selecting values for j, k, and M. For instance, consider a memory with a datawidth of 64 bits and a depth of 128 locations, resulting in a memory of 8Kbits. The address A has a length of 7 bits, where {A0, A1} index the columns, and {A2, A3, A4, A5, A6} index the rows. Here, the memory matrix M has dimensions of 2×4, and C consists of 16 copies of M.

The column mux ratio, denoted by m, is a critical parameter in memory arrays as it determines the number of memory cells connected to a shared bit line. Selecting an appropriate value for m involves a trade-off between memory density, access time, and power consumption. A higher column mux ratio affects the aspect ratio and memory matrix M. However, this also leads to larger capacitance and longer bit lines, causing slower access times and potentially higher power consumption. In contrast, a lower column mux ratio enhances access time. A series of tests were conducted using the foundry compiler to generate multiple memory IPs with varying speeds and densities before designing the SRAM-based PUF. Memories with sizes of 1kbytes and 4kbytes were selected from the results. The memories with a bitcell size of approximately $\sim 0.65\,\mu m^2$[1] are labeled as low-density and high-speed, while those with a bitcell size of approximately $\sim 0.50\,\mu m^2$ are labeled as high-density but low-speed. This process helped to choose the most appropriate memories for the design.

A.1.2 Design and Evaluation of SRAM-Based PUFs

As depicted in Fig. A.1, hASIC requires a single SRAM-based PUF to generate the secret key. After selecting suitable representative SRAMs for SRAM-based PUFs, the next step is to find an appropriate and robust SRAM-based PUF for hASICs. In this regard, the investigation has focused on studying the impact of design-time decisions on the effectiveness and quality of SRAM-based PUFs. A chip was designed using a 65 nm commercial PDK to assess the impact of various memory- and chip-level parameters. The chip featured eleven SRAM macros, which were comprehensively evaluated for performance. The SRAM compiler considered several parameters at the memory level, such as the number of addresses, words, aspect ratio, and bitcell design. Furthermore, during the floorplan phase, chip-level decisions were made concerning the placement, rotation, and power delivery strategy of each SRAM macro within the testchip. All of these factors were taken into account for the evaluation. The study analyzed 50 fabricated chips through physical measurements to assess the reliability, bias pattern, entropy, uniqueness, and randomness of different SRAM configurations.

[1] The SRAM IPs are generated by a major foundry's compiler and are considered foundry IPs. Further details are omitted.

A.1.2.1 Silicon Demonstration

The primary objective in creating silicon is to demonstrate the assessment of SRAM-based PUFs. A total of 11 SRAMs with different possible orientations have been utilized in the test chip, as depicted in Fig. A.2. The initial orientation is $R0$, which represents a rotation angle of zero degrees. The abbreviation MX indicates a mirroring process along the x-axis. The symbol R270 indicates a rotation of 270°, while $R90$ denotes a 90-degree rotation. $MY90$ signifies a mirroring process along the y-axis, followed by a 90-degree rotation (all rotations are in the anti-clockwise direction). Figure A.4 illustrates the placement of SRAMs and their respective orientations inside the chip. The chip includes six separate SRAM-based PUFs, some replicas identified by underscored letters (a, b, c). This results in eleven SRAM-based PUFs within the chip.[2]

The chip has a simple serial interface and eleven different SRAM-based PUFs. All SRAM memories are single-port and use six transistors per bitcell. Additionally, two distinct types of memories were utilized. A streamlined architecture allows for seamless data acquisition across several SRAM-PUFs. A data vector is transmitted via the serial interface using the *shift_in* input, while *shift_in*, *shift_enable*, and *data_out_enable* serve as control bits. The serial input reads one bit of the data vector during each clock cycle. After the data vector read operation, the shift register accumulates the entire data vector. The serial interface utilizes 15 bits for addressing and selecting eleven SRAM-based PUFs, with 10 bits for accessing the address of the SRAM-PUF.[3] To select the PUF, 4 bits are needed, along with an additional bit to enable the read operation. After loading the shift register, the data is dispersed to memory selection and address block. At the output of the serial interface, 70 bits are retrieved, including 64 bits of data, as well as three start bits and three stop bits. For smaller data widths, such as 1024 × 32, the length of the data vector is still 64 bits to

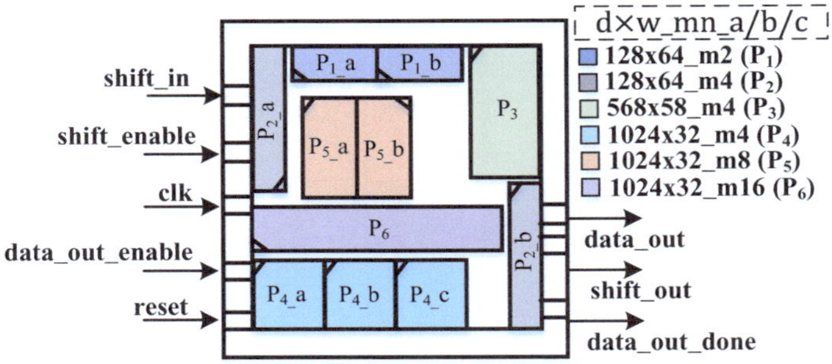

Fig. A.4 The simplified architecture of the SRAM-based PUFs

[2] d denotes the depth or number of addresses, w denotes the datawidth, n denotes the number for ratio, and m denotes the column mux ratio.

[3] The maximum number of addresses that can be accommodated is 1024, equivalent to 2^{10}. When dealing with addresses less than 1024, the residual address bits should be set to zero.

maintain consistency, but the last 32 bits will be zeros. This will make the read operation more convenient and ensure consistent data-read from the architecture.

The chip comprises fast and slow memories, with smaller ones (128 × 64) categorized as fast. To maximize the area utilization, all memories were manually placed multiple times. To start, the design was synthesized using the commercial tool Cadence Genus. The chip does not require any special constraint for the timing; a target frequency of 8MHz has been maintained to enable sequential data reading without encountering data corruption. Three flavors of the standard cell library (LVT/SVT/HVT) have been utilized to align with the industrial standard. The chip layout was generated using Cadence Innovus for P&R. The design underwent physical compliance verification, including DRC and LVS checks. The control logic inside the chip is minimal, with most of the area occupied by memories. As a result, the number of buffers, combinational cells, inverters, and sequential cells in the circuit is less than 5%, with 233 buffers, 640 combinational cells, 881 inverters, and 54 sequential cells. After sign-off, the layout is generated as a GDSII file, which is then sent to the foundry for fabrication. The chip was fabricated successfully, and bench tests were conducted to gather data. The layout, shown in Fig. A.5, reveals the placement of I/O cells and memories, with a black rectangle included to aid in identifying the lower right corner of the chip. The chip's I/O pins and power stripes are visible, running both horizontally and vertically. The placement of the SRAM-based PUFs is clearly visible, with a yellow color in the left panel of the same figure. Signal routing in the design utilized all metals between M2 and M7. To create a power ring around the core, M5 and M6 were employed. In addition, power distribution across the core was achieved through the use of horizontal and vertical stripes in M8 and M9. The layout also includes the seal ring, die and metal fills to meet the foundry requirements. The chip size is 1 mm^2. To validate the design, 50 chip samples were packaged in a DIP-28 form, all confirmed to be fully functional.

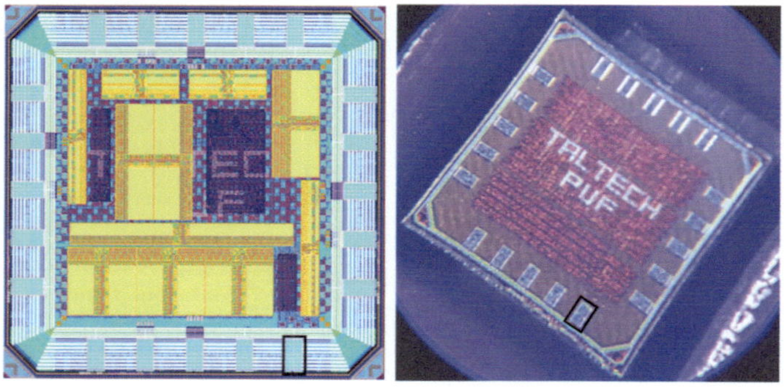

(a) (b)

Fig. A.5 Layout (left) and die micrograph (right) of the fabricated chip. The highlighted pin marks the lower-right corner [3]

A.1.2.2 Testing and Measurement of SRAM-Based PUFs

A custom printed circuit board (PCB) was designed and manufactured for this specific task to evaluate the ASIC prototype. The PCB contains necessary components, such as a DIP-28 socket, relays, and passive components, to facilitate measurements and filter out power supply noise. Figure A.6a illustrates a 2D representation of the PCB layout and component placement. Additionally, Fig. A.6b shows the images of the received packaged chips, visually representing the final product.

Figure A.7 depicts the testing setup for the chip, where the chip is mounted on the PCB. Raspberry Pi 3 Model B was used to control the chip during testing, enabling smooth communication and efficient management of the testing process. The relays on the PCB played a critical role in controlling the power supply, selectively turning on and off the VDD (1.2 V) and VDDIO (3.3 V) power sources, allowing for flexibility and control over the chip's power supply configuration.

In the initial validation phase, the leakage power of all 50 samples was measured. This was done using a picoamp precision ammeter while the chip was idle. The results and distribution of these samples are depicted in Fig. A.8. The distribution's mean value and standard deviation were 6.70 and 2.17, respectively. The typical case was found to be near the mean value. These results were consistent with the expected power reports obtained during the physical implementation for the typical corner at 25 °C with an operating voltage VDD of 1.2 V. The best and worst chips are highlighted in terms of performance. These chips consume the highest and lowest leakage currents, as illustrated in Fig. A.8. The analysis reveals that chip samples do not exhibit a bias towards a specific process corner, but exhibit notable skews between them. This means that process variation significantly impacts the

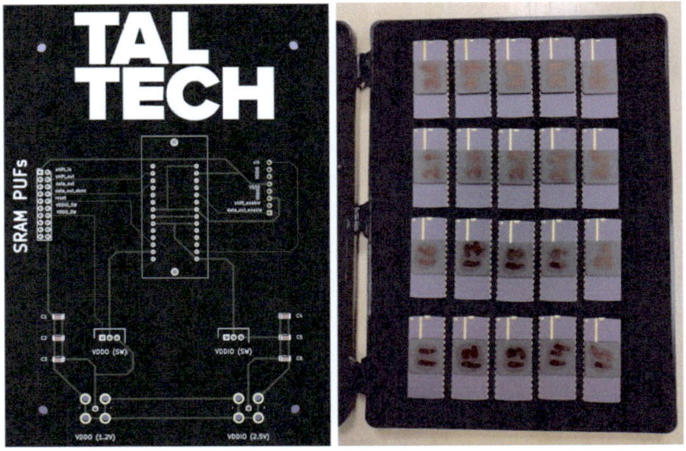

(a) (b)

Fig. A.6 The designed PCB and fabricated chips

Fig. A.7 Testing setup for the chip

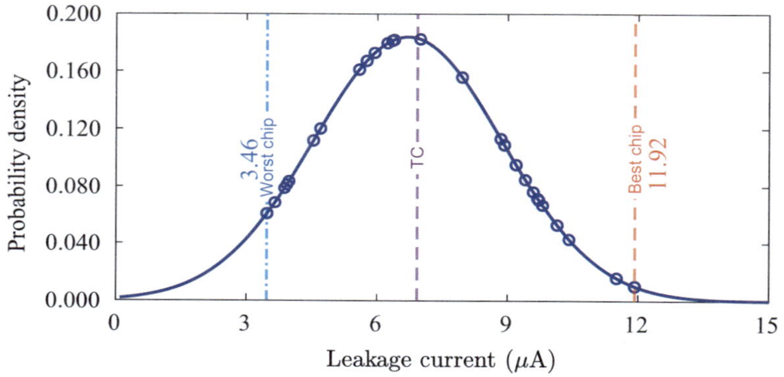

Fig. A.8 The distribution of leakage current across 50 chips and the typical corner (TC) from physical synthesis

samples, which can ultimately influence the behavior of PUFs. During the second phase of the experiments, the responses of the PUF were recorded from the chips by power cycling each chip ten times. It was recognized that relays were important for switching the chip on and off, allowing for a passive power cycle completion. To collect the corresponding PUF response, the Raspberry Pi transmitted serial bits to the chip and stored them in a text file. This process was repeated for subsequent experiments, with the chip being turned on and off. The experiment was repeated ten times to assess the stability of the PUF's response. The total number of bits stored was calculated by multiplying the datawidth of each memory by its respective depth. Four instances of 128×64, six instances of 1024×32, and one instance of 568×58 contain a total of 32.768 K, 196.608 K, and 32.944 K bits, respectively. The total number of bits collected is 262.320 K per chip for a single power cycle. When considering 10 power cycles, this number increases to 2623.2 K bits per chip.

A.1.3 Results and Observations

This section evaluates the robustness of SRAM-based PUFs for various sizes, including reliability, bias pattern, entropy, uniqueness, and randomness. The metric corresponding to each SRAM-based PUF measurement was evaluated using the data from 50 chips. This analysis was performed for all SRAM-based PUFs within a single chip.

A.1.3.1 Robustness Evaluation

The panels in Figs. A.9, A.10, A.11, A.12, A.13, and A.14 illustrate the trends of WCHD, HW, MHW, and BCHD for six distinct SRAM-based PUFs. The x-axis represents the measurement number (power cycles 0-9) and the y-axis shows the corresponding value. The WCHD of all the SRAM-based PUFs is less than 10%, indicating good reliability. Most of the SRAM-based PUFs exhibit a WCHD of approximately 7%, which suggests favorable PUF characteristics for environmental effects. Two chips show a different profile, while the others lie between 5 and 7%. In order to assess the stability of a specific bitcell's response over time, measurements of WCSHD for P_4 and P_5 were taken. The results indicate that the behavior of the two responses appears to be quite close and unaffected by environmental or measurement noise. The trend of WCSHD indicates that the responses exhibit a WCSHD of approximately 6–7% over measurements.

MHW displays the evaluation of entropy for SRAM-based PUFs. MHW is calculated by XORing the startup pattern with the bias pattern and finding the percentage of ones. The results show that MHW for all SRAM-based PUFs is within the range of 0.5 ± 0.1, except for P_1, which exhibits MHW values ranging from 0.38 to 0.62. Although a trend of few chips appears near 0.4, most of the chips were located closer to the ideal value, indicating good entropy. The uniqueness of the PUFs was evaluated in panel (d) of Figs. A.9, A.10, A.11, A.11, A.12, A.13 and A.14, which shows the BCHD for both fast and slow SRAM-based PUFs, with the probability density on the x-axis and the BCHD values on the y-axis. The distributions are centered around 0.5, and the more narrow the distribution, the better and closer to the ideal value. Overall, the PUFs demonstrate good uniqueness. The randomness of the PUFs was analyzed by computing HW, which yielded a fractional value of 0.5 ± 0.3 for all SRAM-based PUFs, indicating good randomness. Table A.1 summarizes the findings. The table's first column lists memory types, while subsequent columns show WCHD, MHW, and BCHD results. The SRAM-based PUF P_4 demonstrated the highest reliability and uniqueness across all 50 chips. Conversely, the SRAM-PUF P_3 showed the highest entropy for all chips. Notably, all SRAM-PUFs exhibited randomness close to the expected 50% value, with most PUFs demonstrating good reliability (over 90%). The variance in entropy highlights that the uniqueness and randomness of all SRAM-PUFs are close to ideal values.

Two additional experiments were conducted to study the behavior of the SRAM-based PUFs placed adjacent to each other in a specific region of the chip. Three identical SRAM-based PUFs (1024×32_m4_a/b/c) were placed in the bottom left corner, as shown in Fig. A.4.

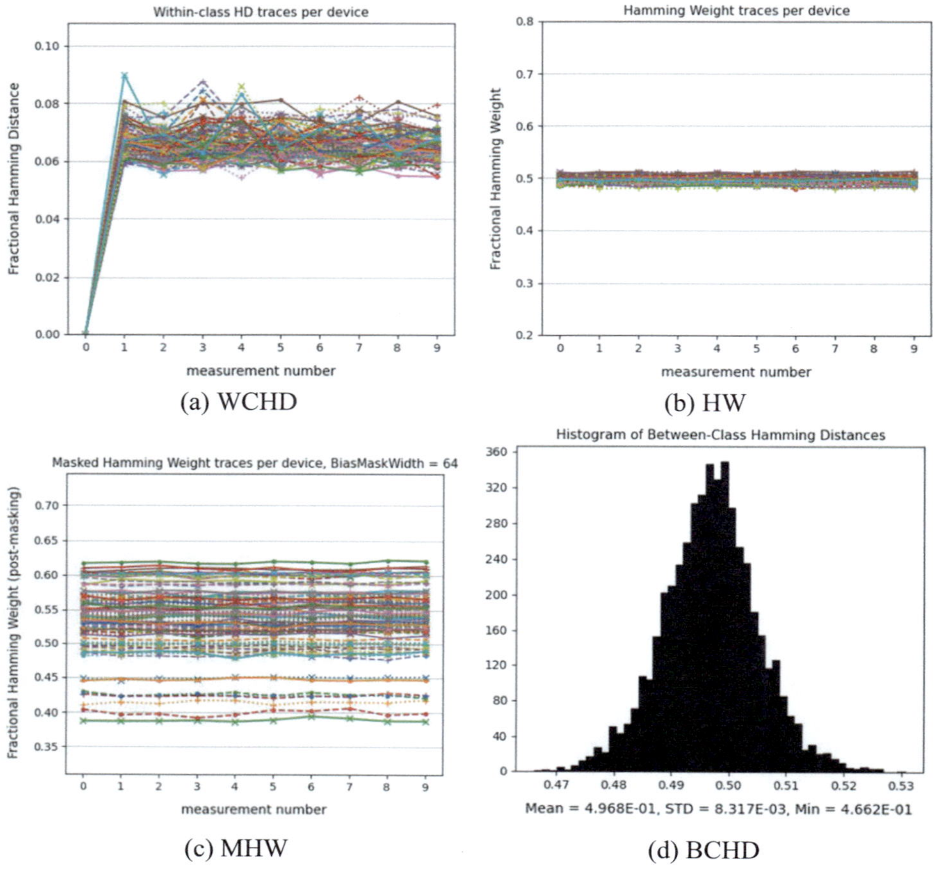

Fig. A.9 The PUF's characteristics of P_1_a and P_1_b

The power mesh of the entire chip was symmetrical and carefully planned for balanced power distribution, except for this region. In the first experiment, the impact of IR drop on the PUF's characteristics was examined, and the results showed that it did not have a measurable effect on the behavior of the SRAM-based PUF. The second experiment analyzed the behavior of the SRAM-based PUF under varying voltage conditions, and the results indicated that the robustness of the SRAM-based PUF was not significantly affected by these voltage conditions. Experiments were also conducted where the two power supplies were turned on at different speeds and in different orders, but there were no measurable changes in SRAM-based PUF quality, due to the chip's power-on-control functionality on its IO cells. This functionality cannot be bypassed, and the core of the chip is only provided power when both VDD and VDDIO are provided. Figure A.4 depicts the placement of the three identical SRAM-based PUFs

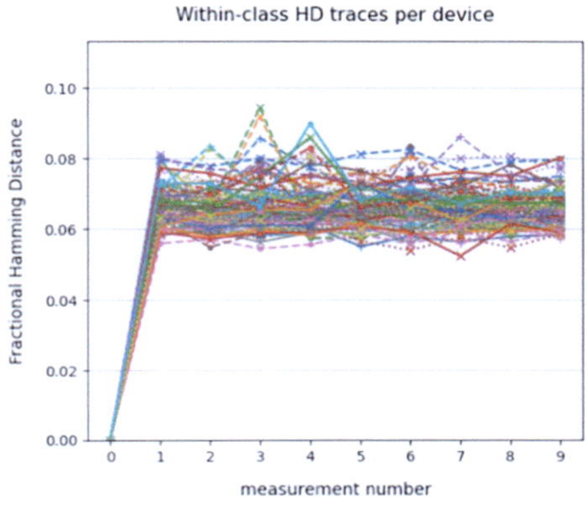

(a) WCHD

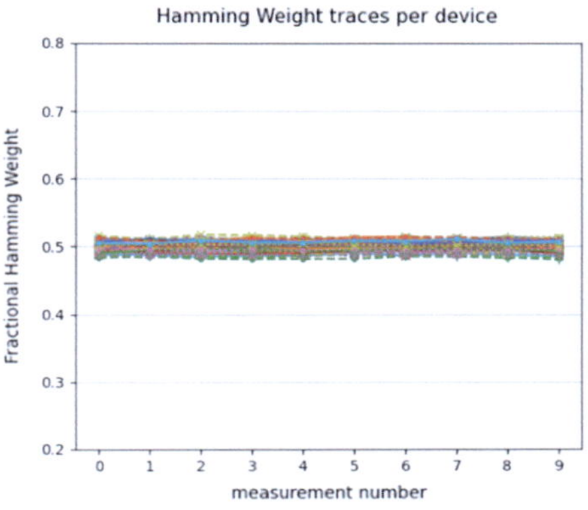

(b) HW

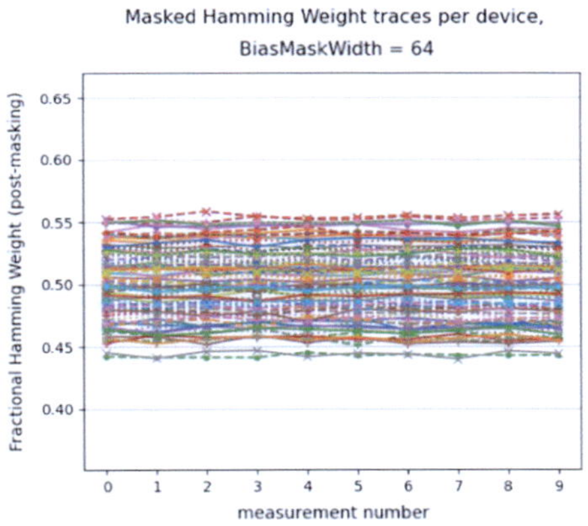

(c) MHW

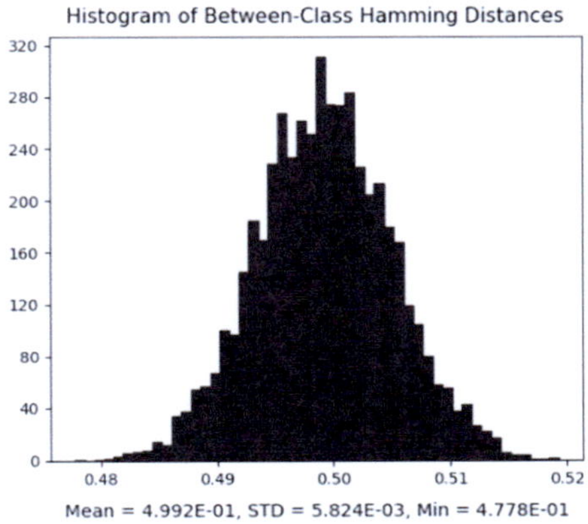

(d) BCHD

Fig. A.10 The PUF's characteristics of P_2_a and P_2_b

Table A.2 provides a summary of state-of-the-art research on SRAM PUFs. In [4], the authors assessed the uniqueness and WCHD of different PUF architectures, including four identical instances of SRAM-based PUFs, on a 65 nm technology node for 192 devices. The findings on WCHD indicated a 95% similarity with similar studies such as [4–9], with the exception of [4], which considered a lower number of chips. In this study, entropy values were consistent with those reported in [8]. While the entropy values in [10] were slightly higher, a fair comparison was difficult due to their reliance on an older technology node. The work demonstrated good uniqueness, similar to previous studies such as [4–10], being very close to the ideal value. The results on randomness were also close to the ideal case and in line with prior research. The purpose of comparing the outcomes is to ensure their consistency with previous research, thereby validating the findings before proceeding with

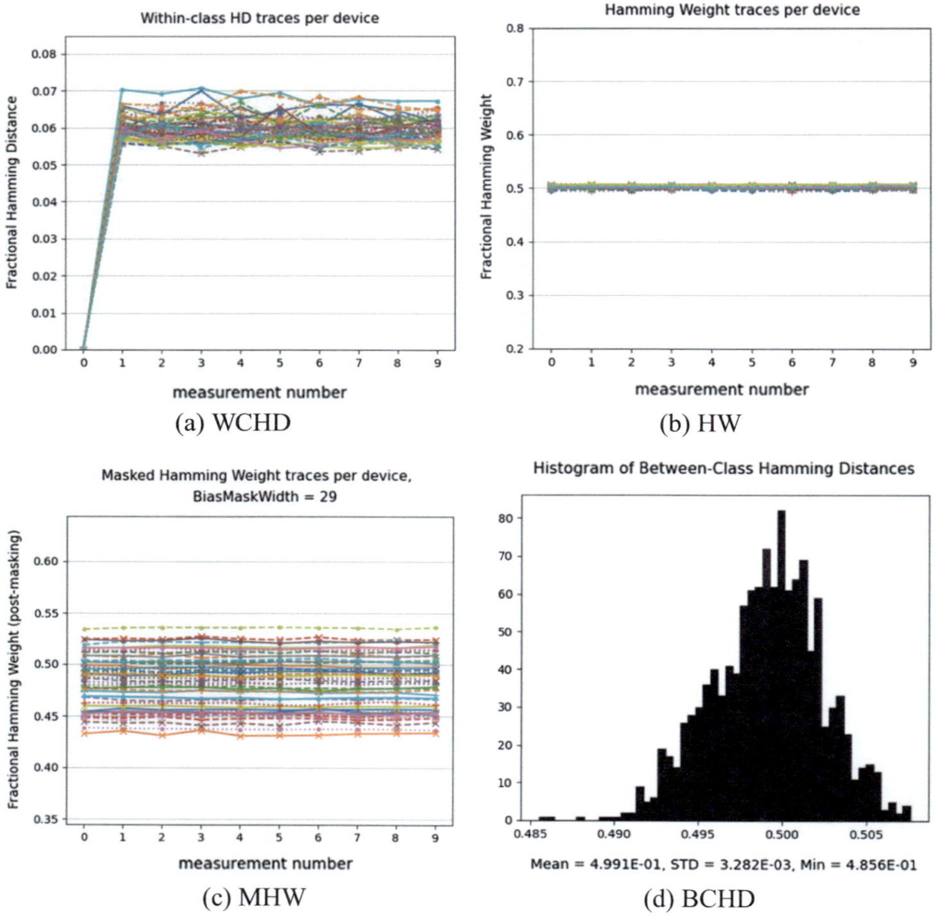

Fig. A.11 The PUF's characteristics of P_3

the next analysis. Overall, the study confirmed that SRAM-PUFs behaved as expected and in line with prior studies.

A.1.3.2 Impact of the Bias Pattern

Next, the analysis focuses on assessing how various factors like sizes, mux selection ratios, memory types, and orientations impact the bias pattern of SRAM-based PUFs. The memory macros are pre-designed, pre-verified memory modules that are crucial to a SoC. They can be embedded in the chip to provide fast and efficient data storage for various tasks. The pins are usually placed on a single edge of the memory. Placing pins along the edges of the memory module makes it easier to connect within the SoC. In this regard, memory

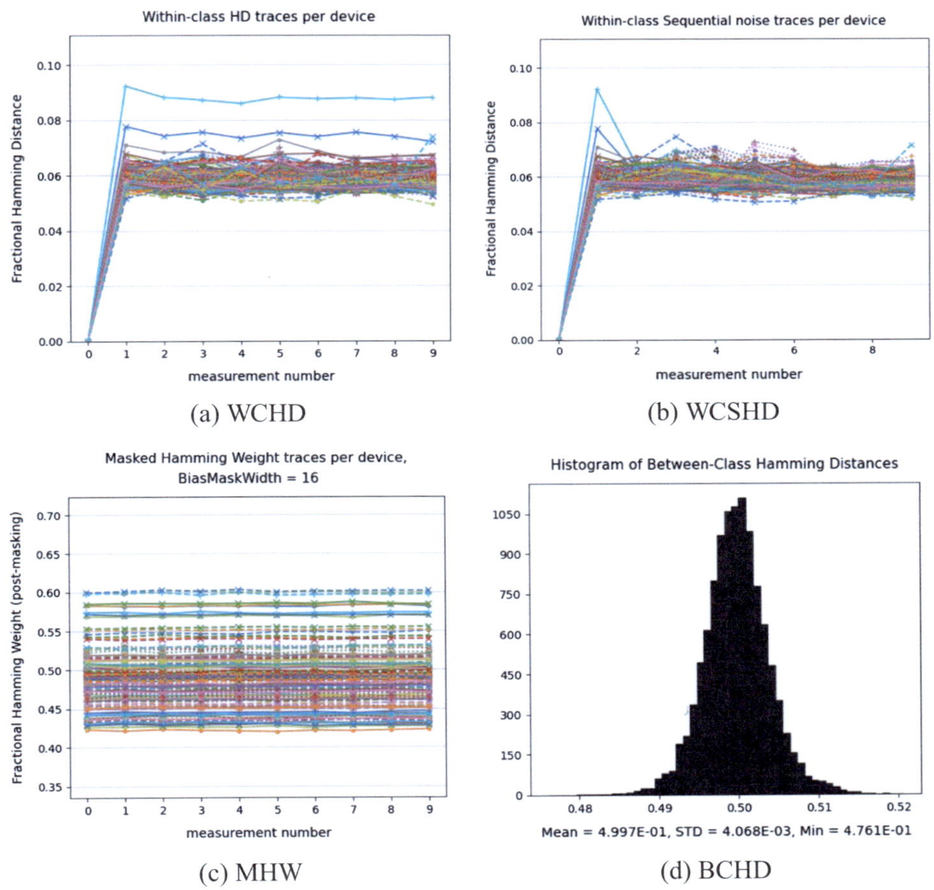

Fig. A.12 The PUF's characteristics of P_4_a and P_4_b

orientation plays a vital role in SoC design to maximize performance, minimize power consumption, and reduce the physical footprint of the chip. Physical designers often rotate and flip memory macros within the SoC design to optimize memory placement and routing. Achieving these objectives requires careful consideration of every aspect of chip layout, including memory organization. The memory orientation on the circuit's floorplan does not affect its functionality but impacts the bias direction (BD).

The start-up pattern of SRAMs P_5_a, P_6, P_2_a, and P_2_b is illustrated in Fig. A.15 to visualize the biasing patterns. The response vectors are concatenated into binary data to determine the bias pattern. Each auto-correlation is computed for each chip, but only on the first measurement. Figure A.16 depicts the correlation between the MHW for all 50 chips. Notably, the direction of correlation varies from one SRAM-based PUF to another when

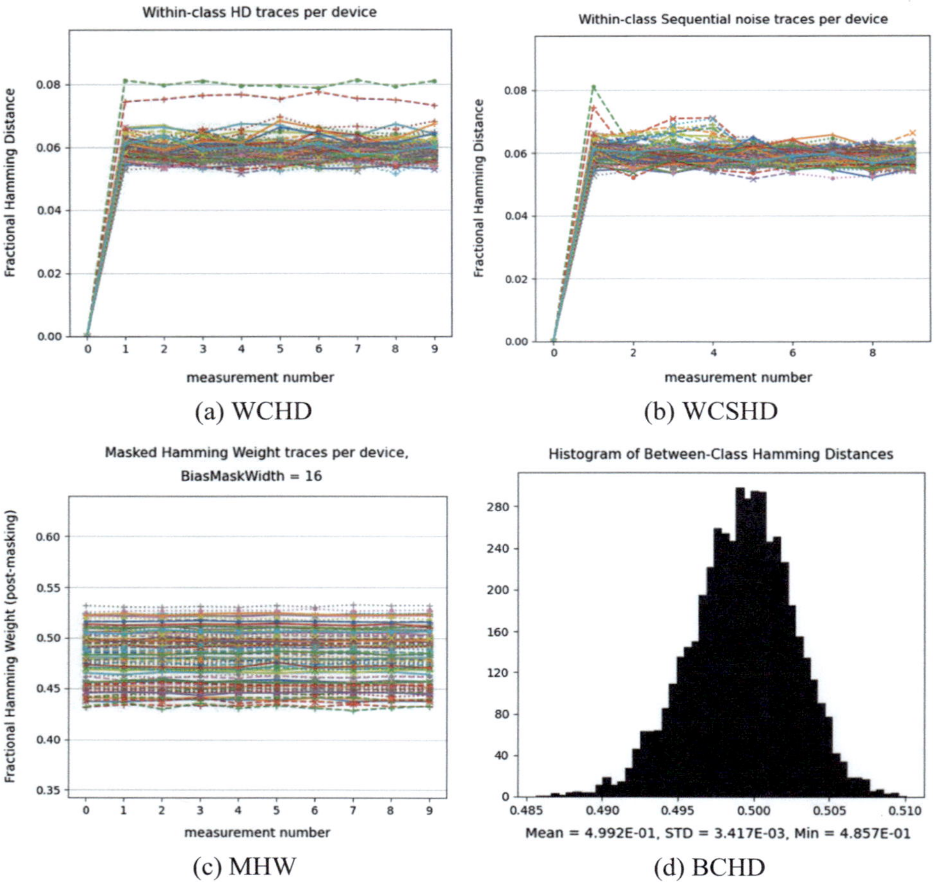

(a) WCHD (b) WCSHD

(c) MHW (d) BCHD

Fig. A.13 The PUF's characteristics of P_5_a and P_5_b

considering P_1_a as the baseline. For instance, the correlation peaks for the baseline and P_2_b are opposite. Thus, the bias direction is reported as negative. Table A.3 summarizes the relationship between memory orientation and bias correlation direction in the last two columns.

Observation 1: Table A.3 presents the data width, bias pattern, and number of instances for various SRAM-based PUF instances. It is important to note that the data width affects the bias pattern, but its direction remains unchanged. For instance, consider two identical memories, P_2_a and P_2_b. Despite their identical nature, each memory exhibits a different bias direction, with one having a positive bias direction and the other a negative bias direction.

Observation 2: The change in mux ratio results in different aspect ratios for SRAM-based PUFs, which in turn causes bias patterns to vary between 1024×32_mux8 and

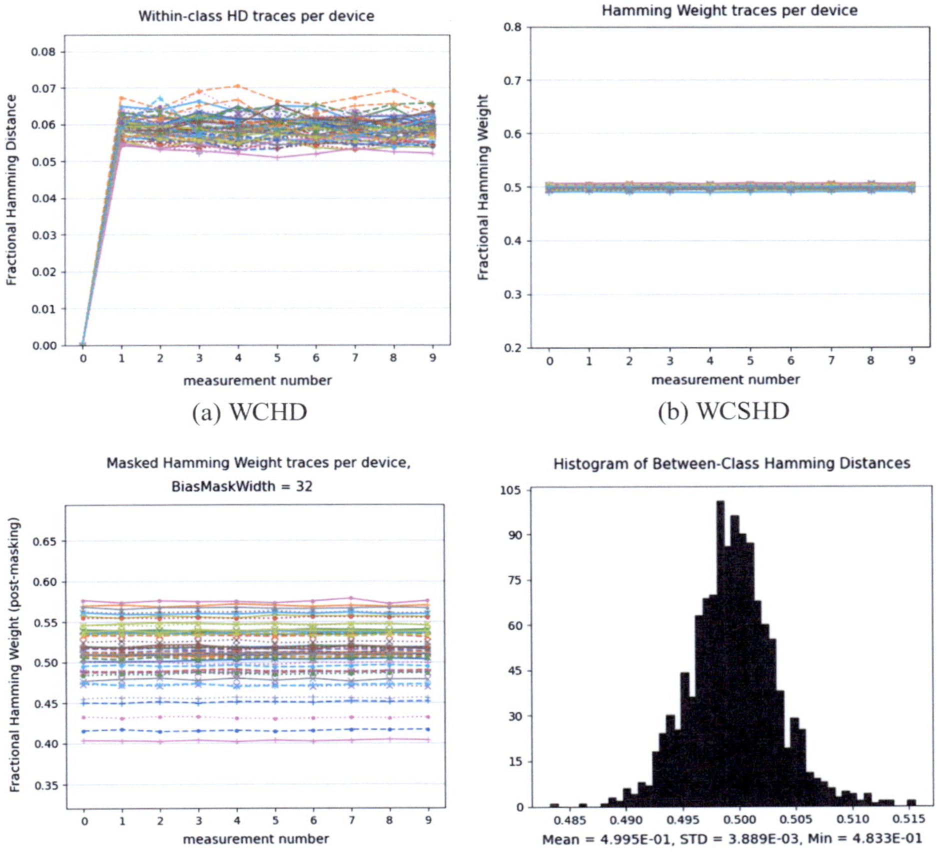

(a) WCHD (b) WCSHD

(c) MHW (d) BCHD

Fig. A.14 The PUF's characteristics of P_6

1024×32_mux16. Thus, the bias pattern is impacted by a larger mux ratio. Identical SRAM-based PUFs exhibit an identical bias pattern. However, the column mux ratio does not affect the direction of the bias pattern.

Observation 3: The width and direction of the bias pattern do not correlate with the fast and slow memories. This is also true for using SRAMs with different sizes, bitcells and column mux ratios. Therefore, it can be concluded that different bias widths or directions cannot be achieved by utilizing different bitcells and column mux ratios.

Observation 4: It should be noted that SRAM-based PUFs can exhibit an alternating bias pattern due to the internal structure of SRAMs. As explained in Sect. A.1.2.1, the SRAM macro comprises two halves: one on the left and the other on the right. This arrangement

Table A.1 Results for the robustness evaluation of SRAM-PUFs

SRAM-PUF	WCHD (%)	MHW	BCHD	Entropy by one-probability
P_1	5.8–8.5	0.391–0.622	0.468–0.526	0.685–1
P_2	5.8–8.8	0.440–0.564	0.479–0.520	0.826–1
P_3	5.5–7.0	0.430–0.539	0.486–0.508	0.811–1
P_4	5.0–9.1	0.435–0.541	0.480–0.519	0.824–1
P_5	5.2–8.0	0.430–0.575	0.486–0.510	0.798–1
P_6	5.1–7.0	0.390–0.580	0.483–0.515	0.713–1

Table A.2 Comparison of results

Ref.	# ICs	Tech. (nm)	WCHD (%)	BCHD	Entropy
[4]	192	65	5–5.5	–	–
[5]	40	65	10	0.489	–
[6] †	11	65	15.98	0.495	–
[7]	1	65	15.42	–	–
[8]	10	65	3.3–5.3	–	0.960–1
[9]	17	90	2–4	0.435	–
This work (P_4)	50	65	5–9.1	0.480–0.519	0.824–1

† Authors evaluated 22000 2-bit cells (each die has 2000 cells)

Table A.3 The biasing pattern of different SRAM-PUF instances

SRAM-PUF	Bias pattern	Orientation	BD
P_1_a, P_1_b	0(32)1(64)0(64), ...	R0, R0	+, +
P_2_a, P_2_b	0(32)1(64)0(64), ...	R90, R270	+, –
P_3	0(29)1(29), ...	R270	–
P_4_a, P_4_b, P_4_c	0(16)1(16), ...	MX, MX, MX	+, +, +
P_5_a, P_5_b	0(16)1(16), ...	R270, MY90	–, –
P_6	0(16)1(32)0(32), ...	R0	+

leads to an interesting observation: the initial 32 bits of P_1_a and P_1_b tend to skew towards zero, followed by an alternating pattern of 64 bits.

Observation 5: The study has confirmed that two memory orientations, namely R270 and MY90, exhibit a negative biasing direction that is distinct from the other memories. This bias direction is independent of various memory attributes such as speed, column mux ratio, size, and utilization of SRAMs with different bitcells.

Observation 6: The direction of bias pattern in SRAM-based PUFs remains unaffected by power planning and overall floorplan of the chip, except for factors related to orientation.

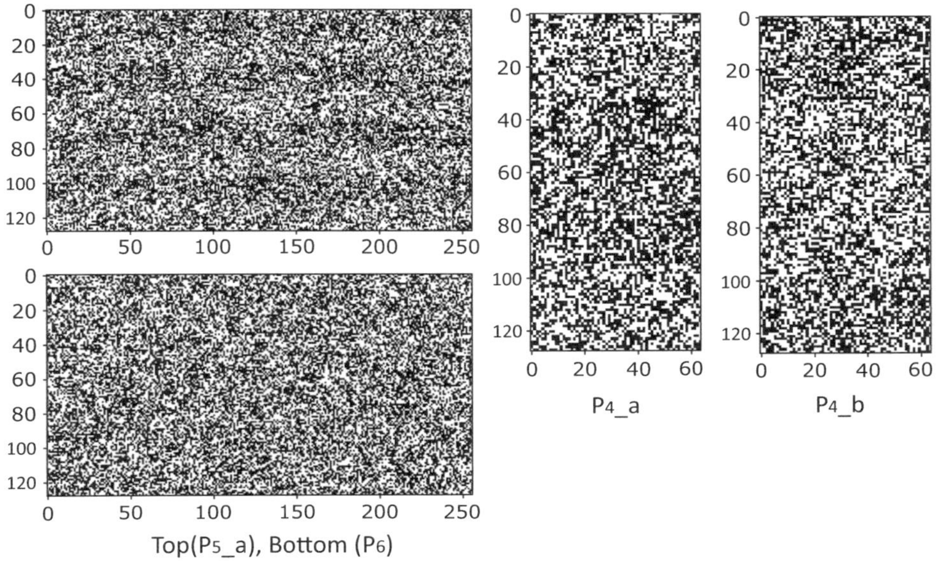

Fig. A.15 The start-up pattern of the P_5_a, P_6, P_4_a and P_4_b SRAM-based PUFs is represented by a sequence of bits, with white spaces denoting a logical one

Observation 7: The intra-die process variation manifests as uniqueness among individual dies, and therefore, it does not affect the bias pattern's orientation. Additionally, all chips originate from a single wafer and have not undergone rotation on the multi-project wafer (MPW) reticle.

Based on observations 1–7, it has been concluded that the direction of the bias pattern remains unaffected by changes in sizes, mux ratios, memory types, memory structure, or process variations. The orientation of the pattern is solely dependent on specific factors. With these observations in mind, a hypothesis was developed to validate the direction of the bias pattern.

Hypothesis 1: The effect observed is due to the orientation of the bitcells themselves. In the R90 orientation of the SRAM, the bitcells are positioned vertically compared to the R0 orientation. When the R90 orientation is taken as a reference, the R270 and MY90 orientations flip the left and right bitcell arrays, as shown in Fig. A.2. This implies that the direction of the bitcell placement associated with the first address of the SRAM gets reversed in these orientations. In simpler terms, the orientation of the bitcell placement corresponding to the first address of the SRAM gets reversed in these orientations.

Hypothesis 2: Analyzing and mitigating the effects of doping variations in SRAM-based PUFs requires a comprehensive understanding of their impact on the bias pattern. The doping of transistors during the lithography process affects their behavior, and variations can occur due to fabrication equipment limitations [11]. The machine moves along the x-axis, doping

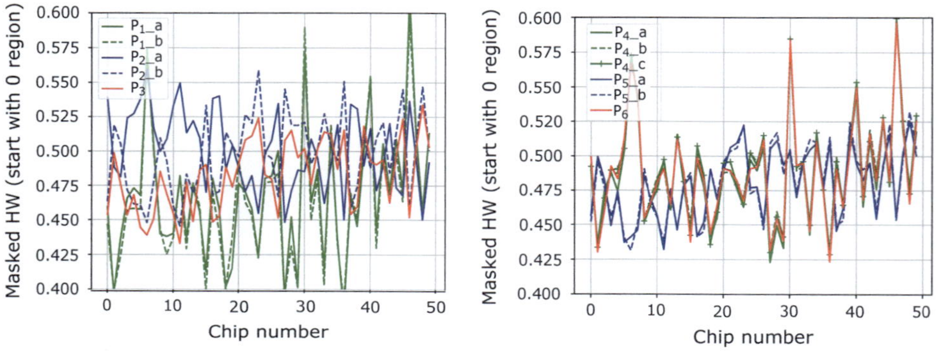

Fig. A.16 The correlation of SRAM-based PUFs for the bias pattern with baseline SRAM-based PUF (P_1_a) [3]

the transistors from left to right or right to left, then steps up until all transistors are treated. These doping variations impact the SRAM cells' initial state and stability, evaluated using the static noise margin (SNM) concept, which represents the minimum noise voltage required to flip the state of the bitcell. The width-to-length (W/L) ratios of the load and access transistors are set to be as close to 1.0 as possible, while the cell ratio determines the cell's stability and size [11]. Doping variation directly affects the W/L ratio, potentially leading to variations in the SNM value. Transistors on the vertical axis experience more significant variations than adjacent transistors on the x-axis. All SRAM-based PUFs are affected by doping variations, but certain orientations exhibit a distinctive negative bias pattern, such as R270 and MY90, where the transistor arrangement becomes reversed or opposite to the baseline SRAM-based PUF's orientation.

Regarding Fig. A.1, the error correction block is placed on the other chip which is executed in the trusted environment. There are two phases in error correction techniques for PUF: enrollment and reconstruction [12]. The PUF response is converted into a codeword during enrollment using an error correcting code [13,14]. The mapping information is stored in the helper data, which is designed not to leak any information about the key. In the encryption scheme, the error correction IP contains the helper data necessary for key reconstruction. However, any modifications to the helper data, whether malicious or not, will prevent key reconstruction. Additionally, the helper data is only valid for the chip on which it was created [2]. During the reconstruction phase, the hASIC performs a new noisy PUF measurement and extracts the PUF key (without noise) from the helper data and the new PUF response. Figure A.1 presents a scheme that explains the complete encryption and decryption of a bitstream. The analysis of SRAM-based PUFs suggests a way to select the most robust SRAM-based PUF from the various ones included in this study. Generating a key from these PUFs provides a reliable and cost-effective solution. While any SRAM can be used as a PUF, the designer should consider using the most robust SRAM-based PUF to generate a secret key.

References

1. Q. Audrey, "A simple guide to AES 256-bit encryption," last accessed on Sep 16, 2023. Available at: https://www.azeusconvene.com/articles/a-simple-guide-to-aes-256-bit-encryption.
2. J.-P. Linnartz and P. Tuyls, "New shielding functions to enhance privacy and prevent misuse of biometric templates," in *Audio- and Video-Based Biometric Person Authentication* (J. Kittler and M. S. Nixon, eds.), (Berlin, Heidelberg), pp. 393–402, Springer Berlin Heidelberg, 2003.
3. Z. U. Abideen, R. Wang, T. D. Perez, G.-J. Schrijen, and S. Pagliarini, "Impact of orientation on the bias of SRAM-based pufs," *IEEE Design & Test*, pp. 1–1, 2023.
4. R. Maes, V. Rozic, I. Verbauwhede, P. Koeberl, E. van der Sluis, and V. van der Leest, "Experimental evaluation of physically unclonable functions in 65 nm CMOS," in *2012 Proceedings of the ESSCIRC (ESSCIRC)*, pp. 486–489, 2012.
5. S. Baek, G.-H. Yu, J. Kim, C. T. Ngo, J. K. Eshraghian, and J.-P. Hong, "A reconfigurable SRAM based CMOS PUF with challenge to response pairs," *IEEE Access*, vol. 9, pp. 79947–79960, 2021.
6. Y. Shifman, A. Miller, Y. Weizmann, and J. Shor, "A 2 bit/cell tilting sram-based PUF with a BER of 3.1e-10 and an energy of 21 fj/bit in 65nm," *IEEE Open Journal of Circuits and Systems*, vol. 1, pp. 205–217, 2020.
7. A. B. Alvarez, W. Zhao, and M. Alioto, "Static physically unclonable functions for secure chip identification with 1.9-5.8% native bit instability at 0.6-1 v and 15 fj/bit in 65 nm," *IEEE Journal of Solid-State Circuits*, vol. 51, no. 3, pp. 763–775, 2016.
8. G.-J. Schrijen and V. van der Leest, "Comparative analysis of SRAM memories used as PUF primitives," in *2012 Design, Automation & Test in Europe Conference & Exhibition (DATE)*, pp. 1319–1324, 2012.
9. G. Selimis, M. Konijnenburg, M. Ashouei, J. Huisken, H. de Groot, V. van der Leest, G.-J. Schrijen, M. van Hulst, and P. Tuyls, "Evaluation of 90nm 6t-sram as physical unclonable function for secure key generation in wireless sensor nodes," in *2011 IEEE International Symposium of Circuits and Systems (ISCAS)*, pp. 567–570, 2011.
10. R. Wang, G. Selimis, R. Maes, and S. Goossens, "Long-term continuous assessment of SRAM PUF and source of random numbers," in *2020 Design, Automation & Test in Europe Conference & Exhibition (DATE)*, pp. 7–12, 2020.
11. B. Cheng, S. Roy, and A. Asenov, "The impact of random doping effects on CMOS SRAM cell," in *Proceedings of the 30th ESSCIRC*, pp. 219–222, 2004.
12. Intrisic ID, "SRAM PUF: The secure silicon fingerprint," last accessed on Sep 11, 2023. Available at: https://www.intrinsic-id.com/wp-content/uploads/2023/03/2023-03-09-White-Paper-SRAM-PUF-The-Secure-Silicon-Fingerprint.pdf.
13. J. Delvaux, D. Gu, D. Schellekens, and I. Verbauwhede, "Helper data algorithms for PUF-based key generation: Overview and analysis," *IEEE Transactions on Computer-Aided Design of Integrated Circuits and Systems*, vol. 34, no. 6, pp. 889–902, 2015.
14. R. Maes, P. Tuyls, and I. Verbauwhede, "A soft decision helper data algorithm for SRAM PUFs," in *2009 IEEE International Symposium on Information Theory*, pp. 2101–2105, 2009.

Glossary

Chip Definition A chip $\text{Chip}_{\text{ReBO}}$ is composed of two main parts: a section implemented with standard cells and an eFPGA.

Standard Cells standard cells are fixed-function logic elements used in the chip's design and are not reconfigurable after manufacturing.

eFPGA Definition An eFPGA is a reconfigurable hardware block embedded within the chip, containing an array of CLBs, interconnects, and I/O blocks.

LUT Definition A LUT within the eFPGA is a reconfigurable element that can implement any logic function of n inputs, where $n \leq 6$. The LUT has 2^n configuration bits.

Configuration A configuration is a specific arrangement of logic and routing resources in the eFPGA, typically defined by a bitstream.

Logic Function A logic function is a mapping $f : \{0, 1\}^n \rightarrow \{0, 1\}$ for some integer n.

Z. U. Abideen and S. Pagliarini, *Reconfigurable Obfuscation Techniques for the IC Supply Chain*, Synthesis Lectures on Digital Circuits & Systems,
https://doi.org/10.1007/978-3-031-77509-3

MIX
Papier aus verantwortungsvollen Quellen
Paper from responsible sources
FSC® C105338

If you have any concerns about our products,
you can contact us on
ProductSafety@springernature.com

In case Publisher is established outside the EU,
the EU authorized representative is:
Springer Nature Customer Service Center GmbH
Europaplatz 3, 69115 Heidelberg, Germany

Printed by Libri Plureos GmbH
in Hamburg, Germany